Advances in Experimental Medicine and Biology

Volume 1173

Advances in Experimental Medicine and Biology presents multidisciplinary and dynamic findings in the broad fields of experimental medicine and biology. The wide variety in topics it presents offers readers multiple perspectives on a variety of disciplines including neuroscience, microbiology, immunology, biochemistry, biomedical engineering and cancer research.

Advances in Experimental Medicine and Biology has been publishing exceptional works in the field for over 30 years and is indexed in Medline, Scopus, EMBASE, BIOSIS, Biological Abstracts, CSA, Biological Sciences and Living Resources (ASFA-1), and Biological Sciences. The series also provides scientists with up to date information on emerging topics and techniques. 2018 Impact Factor: 2.126

Content published in this book series is peer reviewed.

More information about this series at http://www.springer.com/series/5584

Yan-Zhong Chang
Editor

Brain Iron Metabolism and CNS Diseases

Springer

Editor
Yan-Zhong Chang
College of Life Sciences
Hebei Normal University
Shijiazhuang, Hebei, China

ISSN 0065-2598 ISSN 2214-8019 (electronic)
Advances in Experimental Medicine and Biology
ISBN 978-981-13-9591-8 ISBN 978-981-13-9589-5 (eBook)
https://doi.org/10.1007/978-981-13-9589-5

This Springer imprint is published by the registered company Springer Nature Singapore Pte Ltd.
The registered company address is: 152 Beach Road, #21-01/04 Gateway East, Singapore 189721, Singapore

Contents

Contributors

Mohammed M. A. Almutairi Department of Pharmacology and Toxicology, School of Pharmacy, University of Kansas, Lawrence, KS, USA;
Department of Pharmacology and Toxicology, School of Pharmacy, King Saud University, Riyadh, Saudi Arabia

Xian-Le Bu Department of Neurology and Centre for Clinical Neuroscience, Daping Hospital, Third Military Medical University, Chongqing, China

Yan-Zhong Chang Laboratory of Molecular Iron Metabolism, Key Laboratory of Animal Physiology, Biochemistry and Molecular Biology of Hebei Province, College of Life Sciences, Hebei Normal University, Shijiazhuang, Hebei, China

Anna Cozzi Division of Neuroscience, San Raffaele Scientific Institute, Milan, Italy

Xixun Du Department of Physiology, Medical College of Qingdao University, Qingdao, China

Yong-Gang Fan College of Life and Health Sciences, Northeastern University, Hunnan District, Shenyang, China

Guofen Gao Laboratory of Molecular Iron Metabolism, College of Life Sciences, Hebei Normal University, Shijiazhuang, Hebei, China

Chuang Guo College of Life and Health Sciences, Northeastern University, Hunnan District, Shenyang, China

Yansu Guo Beijing Geriatric Healthcare Center, Xuanwu Hospital, Capital Medical University, Beijing, China

Yu-jie Guo West China School of Basic Medical Sciences and Forensic Medicine, Sichuan University, Sichuan, China

Hong Jiang Department of Physiology, Medical College of Qingdao University, Qingdao, China

Qian Jiao Department of Physiology, Medical College of Qingdao University, Qingdao, China

Peng Lei Department of Neurology and State Key Laboratory of Biotherapy, West China Hospital, Sichuan University, Sichuan, China;
West China School of Basic Medical Sciences and Forensic Medicine, Sichuan University, Sichuan, China

Sonia Levi Division of Neuroscience, San Raffaele Scientific Institute, Vita-Salute San Raffaele University, Milan, Italy

Jie Li Laboratory of Molecular Iron Metabolism, College of Life Sciences, Hebei Normal University, Shijiazhuang, Hebei, China

Kuanyu Li Jiangsu Key Laboratory of Molecular Medicine, Medical School of Nanjing University, Nanjing, People's Republic of China

Lin-Bo Li College of Life and Health Sciences, Northeastern University, Hunnan District, Shenyang, China

Paolo Santambrogio Division of Neuroscience, San Raffaele Scientific Institute, Milan, Italy

Honglian Shi Department of Pharmacology and Toxicology, School of Pharmacy, University of Kansas, Lawrence, KS, USA

Limin Shi Department of Physiology, Medical College of Qingdao University, Qingdao, China

Ning Song Department of Physiology, Medical College of Qingdao University, Qingdao, China

Anand Thirupathi Laboratory of Molecular Iron Metabolism, Key Laboratory of Animal Physiology, Biochemistry and Molecular Biology of Hebei Province, College of Life Sciences, Hebei Normal University, Shijiazhuang, Hebei, China

Qing-zhang Tuo Department of Neurology and State Key Laboratory of Biotherapy, West China Hospital, Sichuan University, Sichuan, China

Tao Wang College of Life and Health Sciences, Northeastern University, Hunnan District, Shenyang, China

Yang Xiang Department of Neurology, Chengdu Military General Hospital, Chengdu, China

Huan Xiong Department of Neurology and State Key Laboratory of Biotherapy, West China Hospital, Sichuan University, Sichuan, China

Grace Xu Department of Anesthesiology, School of Medicine, University of Kansas, Kansas City, KS, USA

Shuang-Feng Xu College of Life and Health Sciences, Northeastern University, Hunnan District, Shenyang, China

Peng Yu Laboratory of Molecular Iron Metabolism, Key Laboratory of Animal Physiology, Biochemistry and Molecular Biology of Hebei Province, College of Life Sciences, Hebei Normal University, Shijiazhuang, Hebei, China

Yating Zhang Laboratory of Molecular Iron Metabolism, College of Life Sciences, Hebei Normal University, Shijiazhuang, Hebei, China

Chapter 1
Brain Iron Metabolism and CNS Diseases

Anand Thirupathi and Yan-Zhong Chang

Abstract Iron is the most abundant trace element in the human body. It is well known that iron is an important component of hemoglobin involved in the transport of oxygen. As a component of various enzymes, it participates in the tricarboxylic acid cycle and oxidative phosphorylation. Iron in the nervous system is also involved in the metabolism of catecholamine neurotransmitters and is involved in the formation of myelin. Therefore, iron metabolism needs to be strictly regulated. Previous studies have shown that iron deficiency in the brain during infants and young children causes mental retardation, such as delayed development of language and body balance, and psychomotor disorders. However, if the iron is excessively deposited in the aged brain, it is closely related to the occurrence of various neurodegenerative diseases, such as Alzheimer's disease, Parkinson's disease, and Friedreich's ataxia. Therefore, it is important to fully study and understand the mechanism of brain iron metabolism and its regulation. On this basis, exploring the relationship between brain iron regulation and the occurrence of nervous system diseases and discovering new therapeutic targets related to iron metabolism have important significance for breaking through the limitation of prevention and treatment of nervous system diseases. This review discusses the complete research history of iron and its significant role in the pathogenesis of the central nervous system (CNS) diseases.

Keywords Iron metabolism · Neurodegenerative diseases · Oxidative stress · Alzheimer's disease · Parkinson's disease

Iron is an essential player of many biological processes in the brain such as DNA synthesis, oxygen transport, myelin formation, and mitochondrial functions. It is necessary to maintain the iron homeostasis for normal physiological actions of the

A. Thirupathi · Y.-Z. Chang (✉)
Laboratory of Molecular Iron Metabolism, Key Laboratory of Animal Physiology, Biochemistry and Molecular Biology of Hebei Province, College of Life Sciences, Hebei Normal University, 20, Nan Er Huan Eastern Road, Shijiazhuang 050024, Hebei Province, China
e-mail: chang7676@163.com

A. Thirupathi
e-mail: ananthzeal@gmail.com

Y.-Z. Chang (ed.), *Brain Iron Metabolism and CNS Diseases*,
Advances in Experimental Medicine and Biology 1173,
https://doi.org/10.1007/978-981-13-9589-5_1

brain. However, misregulation of iron homeostasis can cause neurodevelopmental and neurodegenerative diseases through different mechanisms. Homeostatic mechanism provides the optimum condition to the cells to carry out some essential functions including maintaining the equilibrium status of the iron concentrations within the cells and buffering the molecules for stopping the toxic buildup within the cellular compartments. When the iron level is increased over the iron sequestration capacity, the concentration of the iron can increase, which can cause several damages in the cellular organelles and cell death in the brain. However, the iron buildup with neurodegeneration is a primary event or secondary event which is unknown. Here, we discuss the history of iron research and iron metabolism in the brain. Next, we discuss the disturbances in the iron homeostasis causing neurodegenerative diseases, and finally, we discuss the important findings that can overcome the iron accumulation causing diseases and its demerits.

1.1 The Evolution of Iron Research

After the bronze era, Iron Age (~1200–1000 BC) began to dominate every creature in the universe, and it turned into an imperative player of entangled biochemistry, which means iron is indispensable for life. Iron is one of the most superabundant metals that contribute to make earth habitable by generating magnetosphere circling to the planet. Indeed, oxygen-free states of the earth made possible to occur a number of chemical reactions that initiate the life existence. Importantly, the essential compound that possibly creates life is a soluble form of iron, hydrogen, and sulfide from ores. In the earlier life formation, the soluble iron was rich, but it began to reduce due to oxygen addition that means life formation is affected slowly but steadily. Günter Wächtershäuser proposed that chemistry of life occurred in the mineral sources [60, 61]. In the sixteenth century, Nicholas Monarde used the iron for various ailments such as alopecia, acne, wounds, hemorrhoids, and gout, revealed the biological significance of iron [42].

The underlying biological importance was proposed initially by Lemmery and Geofgroy in 1713 [57]. Nonetheless, it was nearly required two centuries to comprehend iron metabolism in the body. Vincenzo Menghini was discovered the presence of iron in the blood at the earlier of seventeenth century in the human body [4]. This group was scorched the blood from various models including birds and humans for searching iron in the blood. In the eighteenth century, Louis Rene Lecanu discovered that RBC has a protein called globulin and red coloring matter called hematosine that contained iron [34, 35]. In the earlier nineteenth century, Fontés and Thivolle recommended an imperative revelation that a body has unique form of iron in relation to the hemoglobin in the serum [18]. They additionally found that inadequacy of iron can affect the serum iron concentration. In the mid of nineteenth century, Holmberg and Laurell found the iron-binding proteins called transferrin (Tf) [26]. In the meantime, Schade and Caroline determined the antibacterial effect of iron-binding proteins, and they proposed that siderophilin, a bacterial iron-absorbing compound, which was indistinguishable to Tf proteins [52]. This finding has crucial information since most of the drugs that are currently used for chelation therapy belong to siderophilins. Henry John Horstman Fenton proposed that iron can become toxic in certain circumstances when the level of iron is surpassing in the cell [14, 15], and the concept was not acknowledged by the researchers until the disclosure of superoxide dismutase which converts superoxide into oxygen and hydrogen peroxide.

Studies from 1937 by John Livingood, Fred Fairbrother, and Glenn Seaborg found an exceptional radioactive labeled iron for analyzing iron homeostasis. Based on this radioactive iron, McCance and Widdowson in 1938 proposed that iron is not excreted from the body [44]. Likewise, they theorized that only a small amount of iron was assimilated from the whole intestinal tract, but the main site of assimilation occurred at the upper locale of the intestine. Their investigation was accurate with respect to iron homeostasis except their suggestion that "in hemochromatosis, the iron deposited in the cells is insoluble and inert, and takes no part in the general equilibrium. Hence there is a true deficiency of free iron, as evidenced first, by the anemia, and secondly, by the cure of the anemia by the administration of iron and for their estimation that the red cell life span was only 30 days." Ferrokinetic studies by Pollycove and Mortimer in 1961 were proposed a clear and surmised size of each iron pool [49].

In 1997, Hediger's group and Andrews' group discovered an important protein called divalent metal protein 1 (DMT1) also known as DCT1 and Nramp2 which transports the dietary iron using H^+ electrochemical gradient as driving force. They found that the expression of DMT1 in the rat increased the uptake of $^{55}Fe^{2+}$ and Fe^{2+}. They also found that the DMT1 expression was reactive to several other divalent metal ions. Vulpe group discovered a multicopper oxidase protein called hephaestin in the small intestine in 1999 and Chen group in 2010 identified another type of multicopper ferroxidase called zyklopen in the placenta, and the expression of these proteins is necessary for iron efflux from placental cells and small intestine [6, 59]. The iron entry into plasma, iron utilization and storage are determined by the interaction of the peptide hormone called hepcidin and ferroportin (FPN1). FPN1 was identified by three different groups using different methods. The first group found FPN1 through zebrafish mutant gene [11]. The second group identified FPN1 as a transcript [39]. This group also identified the Dcytb (duodenal cytochrome b) (also known as Cybrid1) as an iron-regulated protein using a subtractive cloning strategy [40], and the third group identified the FPN1 as an iron-responsive element (IRE)-containing mRNA [55]. The role of hepcidin in relation to iron regulation was identified by Nicolas group in 2001 through the HAMP1 gene deletion, which encodes hepcidin [45]. All of these discoveries have been greatly promoted to understand iron metabolism in the body.

1.2 Iron Metabolism and Cell Iron Homeostasis

1.2.1 Iron Metabolism in the Body

Understanding of iron metabolism begins with dietary iron that can be efficiently taken to the cells through duodenal enterocytes. This process is firmly managed by several sensory on and off mechanisms in the epithelium, which is carried out by a number of proteins such as DMT1, cytochrome b ferrireductase, ferroportin, TfR1 and hephaestin. Several enigmatic concepts existed in earlier studies about iron absorption. One of the concepts proposed by Hahn and Whipple is known as "mucosal block mechanism." They declared that the first dose of given iron is saturated the mucosa with ferritin for blocking the absorption of the second dose [21]. In the 1950s and 1960s, some studies had contended against mucosal block mechanism that anemia can be upregulated by the iron absorption in the iron deficiency condition, and it was reduced by iron overloading condition. Although studies had given a clear idea about iron absorption by bowel in the middle of the twentieth century, the bowel components that direct iron absorption are still elusive. Conrad and Crosby in 1963 demonstrated that the entry of radioactive iron in mucosa sloughed off when the body needs low iron [8, 9]. They inferred that the iron regulation was determined by the exchange of iron between mucosal to blood, and the concept is now widely accepted. However, the mechanism remains elusive.

The intestinal mucosa responds if the body's iron level is altered and facilitates the iron absorption accordingly. Iron deficiency condition increases absorption, whereas iron overload condition decreases the iron absorption. The loss of iron occurs through desquamation process and an adult man losses 1 mg of iron, and this level varies with women during their menstruation periods. The healthy population needs 1–2 mg of iron per day for compensating the loss of iron. Dietary iron absorption takes place in the duodenum and upper jejunum. Several proteins are involved to carry out this process such as DMT1, FPN1, and Dcytb. The dietary irons are mostly as inorganic form, and it is reduced by the Dcytb, a ferrireductase enzyme present in the brush border of microvilli. Then, DMT1 transports the ferrous form of iron from the apical membrane of the duodenal enterocytes into intestinal epithelial cells where it can be stored as ferritin or transported into circulation across the basolateral membrane. FPN1 transports the iron from the basolateral membrane to circulation and non-transported iron is removed from the body through exfoliation of the intestinal epithelium. This epithelium exfoliation may represent the iron excretion because these cells are expressing TfR1 at the basolateral membrane and iron from the plasma can enter through endocytosis. For transporting iron through the basolateral membrane, iron must be transported through cell cytoplasm, but this mechanism is not fully understood. However, this iron transport possibly occurs through cytosolic chaperons and iron-binding proteins or transcytosis. Iron transport from the basolateral membrane of enterocyte to circulation requires the changes of redox state which is carried by the hephaestin in the duodenum and ceruloplasmin (CP) in other parts of the body which converts the intracellular Fe(II) back to extracellular Fe(III) for transferrin [59].

1.2.2 Cellular Iron Homeostasis

Several researchers are involved to explore the cellular level of iron homeostasis. We referenced here some of the remarkable researches regarding iron homeostasis. Finch group first investigated the reticulocytes iron transference in 1949 [17]. In little later, Jandl and Katz groups proposed the presence of TfR1 for take-up irons from the reticulocytes [30]. Evan Morgan and his group were evidenced by the cellular level of iron accession mechanism of Tf in the 1970s [2, 43]. Furthermore, Fineberg and Greenberg provided the molecular level of iron accession mechanism [16]. They proved that the administration of iron increases the ferritin protein synthesis, and this experiment was further carried out by Drysdale and Munro [12, 63]. Zahringer group detailed that iron administration increased the ferritin mRNA content in the actinomycin-treated animals. Munro group explained that iron-sensitive factors called iron-responsive elements (IREs) and iron regulatory proteins (IRPs) can regulate ferritin mRNA translation. They reported that ferritin availability to iron was decreased by 5′ UTR deletion containing a conserved 28 base sequences from ferritin mRNA [33]. Some earlier studies reported that TfR1 and ferritin expression are absolutely inverse. When iron is overabundance, the level of TfR1 is constrained; whereas the iron is less, the level of TfR1 is high. Therefore, ferritin and TfR1 levels are precisely altered at the cellular level. Owen and Kuhn evidenced that the sequence within the 3′ UTRs is needed for iron-dependent TfR1 expression, while TfR1's promoter region is not needed [47]. Further research evidenced that IREs present in ferritin while TfR1s bind with an IRE-binding proteins called IRPs. There are two types of IRPs having almost similar function. Higher iron can increase the translation of ferritin mRNA and decrease the stability of TfR1 mRNA. This crucial mechanism is orchestrated by the presence of IREs at the 5′ UTR of ferritin mRNA and 3′ UTR of the TfR1 mRNA. These IREs of both ferritin and TfR1 can bind with cytosolic proteins IRP1 and IRP2. Cells acquire iron from the plasma Tf. After loading iron to Tf, the complex can bind with TfR1 on the cell surface and undergoes endocytosis through clathrin-coated pits. For triggering the release of iron from the Tf, proton pump induces the acidification of endosome to pH level 5.5. For redox changing to transferring iron by DMT1, ferrireductase STEAP3 reduces Fe^{3+} to Fe^{2+}, and this Fe^{2+} form can be stored in ferritin or utilized by downstream metabolic pathways. After releasing iron from Tf, the Tf is dissociated from the TfR1 and the apo-Tf is entered into circulation where it can recapture the Fe^{3}. Regulation of cellular iron homeostasis is tightly managed by the post-transcriptional mechanism, and especially the IRP-IRE system can mediate this post-translational modification for influencing the synthesis of proteins that are responsible for maintaining the intracellular iron level. In the iron depletion condition of a cell, IRPs bind with IREs in the 5′ UTR or 3′ UTR of mRNA for either suppressing the translation of mRNA such as FTH1 and

FPN1 or by increasing the mRNA translation including TfR1 and DMT1 [24, 25]. Under cellular iron increased conditions, IRPs loses their binding activity to IREs by acquiring iron–sulfur cluster (4Fe–4S cluster) and proteasomal degradation [20, 51].

1.3 Brain Iron Metabolism and Nervous System Diseases

In the brain, the function of iron in DNA synthesis, gene expression, neurotransmission, myelination of neurons, and the mitochondrial system can be crucial. Therefore, the balance of brain iron needs to be strictly regulated. However, iron abnormal metabolism can lead to misbehave many proteins that are associated with brain diseases or disorders. The higher iron deposition or iron deficiency may impair the normal functions of the cells. Over iron accumulation in a specific region of the brain leads to several neurodegenerative diseases such as Alzheimer's disease and Parkinson's disease. Dysregulation of iron homeostasis also sets diseases like Friedreich's ataxia (FA). Understanding iron metabolism in the CNS is increasing interest. In addition, numbers of questions are travelling along with scientific discoveries. For example, what are the characteristics of iron metabolism and its regulatory mechanisms in each cell type of the CNS such as neurons, oligodendrocytes, and microglia and what is the mechanism of iron transport and regulation across the blood–brain barrier?

1.3.1 Iron Homeostasis in the Brain

The brain requires iron for its metabolic process like other organs. However, each type of brain cells such as neurons, oligodendrocytes, and astrocytes possess distinct iron metabolism. In addition, unequivocal concepts exist that how iron cross the blood–brain barrier (BBB) or cerebrospinal fluid (CSF) from the circulation and how it traffics in the brain parenchyma to provide iron to different brain cells. The distribution of iron in the brain is mostly concentrated in the substantia nigra pars compacta (SN), circumventricular organs, globus pallidus, and oligodendrocytes. It is well known that the brain is protected by a special structure called BBB, which can prevent the pathogen and other macromolecular compounds entry into the brain and ventricles. Therefore, transporters and receptors can mediate iron transfer across the luminal membrane of endothelial cells of BBB into the CNS. Nevertheless, these regulatory molecules are associated with many diseases and are the major baffling of current research.

The current understanding of iron homeostasis in the brain begins with the Tf/TfR pathway, may be the major route for iron transport across the luminal membrane of the endothelium. Transferrin-bound iron then be converted into Fe^{3+} and translocated across the endosomal membrane by DMT1. Tf will be returned to the luminal membrane once it releases iron with TfR. Other pathways may also be involved to transport iron across the BBB such as lactoferrin receptor/lactoferrin and GPI-anchored melanotransferrin/soluble melanotransferrin. In addition to these pathways, transferrin-bound iron is transferred across BBB through transcytosis. FPN1 and hephaestin might play a key role in transferring iron across the BBB as Fe^{2+}. Another possible way for iron transfer for across the BBB is carried out by astrocytes. The FPN1 released Fe^{2+} in basolateral surface of endothelial cells. Then, astrocytes oxidize Fe^{2+} into Fe^{3+} by CP through their end-feet processes on the capillary endothelia [38, 46], which are further bound with Tf produced by choroid plexus and oligodendrocytes, suggesting that iron not only enters via endothelial cells of BBB but also enters through epithelial cells of choroid plexus. After entering iron into interstitial fluid or CSF in ventricles, the iron will binds with Tf and diffuses via CSF and interstitial fluid of the brain parenchyma, supply iron to the cells expressing TfR within the CNS. A small amount of transported iron can circulate with citrate and ATP. Different cell types in the brain are involved to acquire iron using different pathways. Neurons acquire iron through Tf, astrocytes acquire iron through DMT1, oligodendrocyte acquires iron from Tf by the TfR1 pathway and from ferritin via TIM-2 receptors, and microglia acquire iron from ferritin. Although several studies are well established the iron homeostasis in the CNS, due to BBB, the iron levels in the peripheral nervous system (PNS) decouple from the brain. However, FPN1 may play the role to export the iron from the CNS to PNS with the help of ferroxidase. Furthermore, PNS has little information regarding iron homeostasis. Since BBB has limited access to iron, it expresses all the iron regulatory proteins as we mentioned above. However, on the contrary, the peripheral nervous system has little information regarding iron regulatory proteins. The individual nerve fiber is protected by the connective tissue matrix like epineurium, perineurium, and endoneurium in the PNS. Unlike BBB, the blood–nerve barrier (BNB) permits the proteins and macromolecules. Both BBB and BNB shared the common feature in transferring iron, such as having enzymes, receptors, and transporters. For example, Tf-bound iron is the main source for CNS, but other sources can also contribute to transfer iron in the CNS. However, PNS depends on Tf–TfR1 pathway for iron acquisition especially Tf is accumulated in the Schwann cell (SC) cytoplasm, suggesting a role of Tf-bound iron in the myelination. Furthermore, SC expresses other iron exporters like FPN1 and ferroxidase CP which associate together to efflux the iron to the axons in the peripheral nerve. CNS has ferritin as an iron source, but PNS does not have a ferritin-dependent active mechanism [23]. PNS acquires iron using NTBI through DMT1 [58].

1.3.2 Iron Entry and Its Metabolism in the CNS

Iron passage into the brain had been a long mystery. Initially, it was believed that the iron entry occurred before the blood–brain barrier maturation at the infant stage. Now, it is known that endothelial TfR1 mediates the brain iron uptake. However, TfR expression is regulated by the availability of iron in CNS. In addition, the lactoferrin receptor (LfR) and lactoferrin, and GPI-anchored melanotransferrin are involved to play a pivotal role in iron transport across the BBB. Non-transferrin-bound iron (NTBI) can also present in the BBB [10]. Blood–brain barrier is consisted of a tight epithelial barrier similar to the duodenum is a place of dietary iron absorption. In addition, the presence of BBB can relatively explain the independence of brain iron from the total body iron. Other sites including retina and testis of the mammalian body which be separated from the systemic circulation, the iron absorption maybe similar to the brain.

Endothelial junctional complexes of BBB and astrocytes can decide the passage of iron into the brain. Endothelial cells of brain express TfR1 in the luminal membrane, and TfR1 complexes with Tf for further iron transferring. Endosomes can internalize these complexes to reduce the ferric to ferrous form by endosomal reductase. Nevertheless, the mechanism by which the iron transports from Tf to brain interstitial system is obscure. There are some possibilities that can bolster this mechanism that is after the iron detachment from the Tf can be siphoned into the cytosol by DMT1. Then, the iron transport across the abluminal membrane into the brain is mediated by the FPN1. Alternatively, TfR1 and Tf release iron in the interface between the endothelial surface and astrocyte end-foot process, which is then oxidized by the CP into the ferric form of iron. Subsequently, it is attached with Tf of brain interstitial fluid (Fig. 1.1). Neurons expressing high levels of TfR1 can utilize efficiently irons from Tf. However, neurons with lesser level of TfR1 expression or failed expression can lead to iron dysregulation causing neurodegeneration.

On the other hand, ferrous iron can be moved as NBTI by either attached with ATP or citrate released from the astrocytes. This NBTI is the primary iron source for the cells which do not express TfR1 such as astrocytes and oligodendrocytes. However, the amount of TfR1's expression is varied in each cell type and its iron status. For example, oligodendrocytes firmly express transferrin, whereas microglia strongly express ferritin. Most of the brain cells acquire iron from the TfR1 and endosomal DMT1 and store iron by ferritin and export via FPN1. It has been reported that endothelium of BBB, neurons, and astrocytes expresses FPN1 [22, 32]. Cerebrospinal fluid can also be the iron entry route for brain across the choroid plexus [50].

Neurons including motor, sensory, and interneurons are linked to regulate the iron metabolism in the brain because neurons are ensheathed by glial cells such as oligodendrocytes, astrocytes, and Schwann cells, and these glial cells are expressing iron regulating transporters, receptors, and exporters such as Tf/TfR and FPN1 for restructuring the neuronal process, myelinate axons, and regulating metabolic functions. Neurons take up iron via Tf/TfR and DMT1. The newly imported iron can be stored as ferritin or exported through FPN1. Then, the exported iron from the neuron is oxidized by astrocytic CP into ferric form [1]. However, within these neurons there are different gene expressions that are linked with iron metabolism, thus understanding of iron export from these neurons to other parts of the brain needs profound clarification.

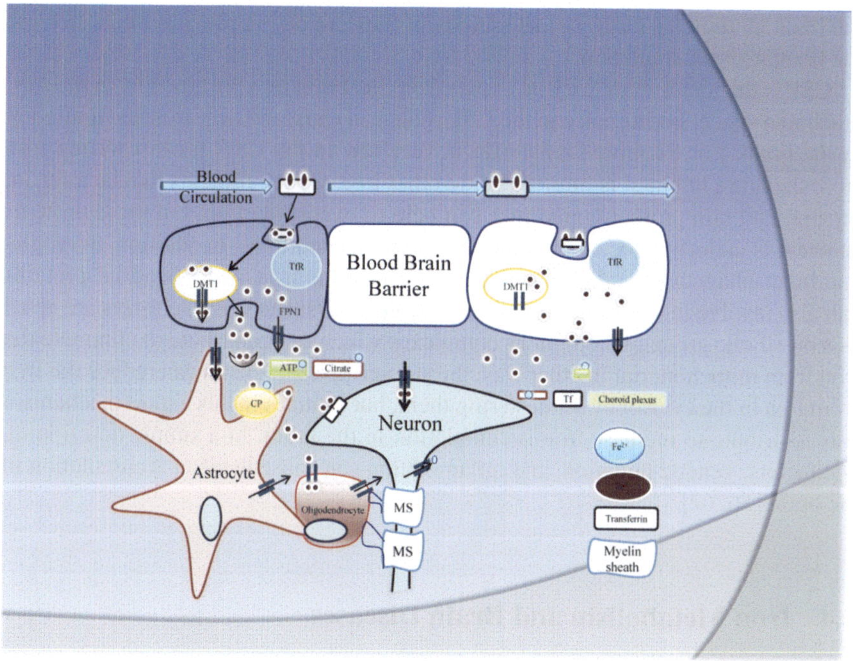

Fig. 1.1 Understanding of iron homeostasis in the brain. TfR1 of endothelial junctional complexes and Tf can transfer the iron with the endosomal process. Detachment of iron from Tf, can be pumped into the cytosol by DMT1 where it can be stored as soluble form of iron. Some iron can be pumped out by FPN1, which is further oxidized by ceruloplasmin into ferric form for Tf. Non-bound iron can be either attached with ATP or citrate which is released from astrocytes

1.3.3 Brain Iron Accumulation

Although brain cells have tight regulation for iron entry, iron is the primary source for several cytosolic proteins and enzymes, mitochondria, and nuclear ferroproteins of brain cells. It has been known that iron accumulation is directly proportional to aging. However, this situation is not connected with pathogenesis. Some special regions such as globus pallidus and substantia nigra in the brain can participate to increase the iron deposition, but factors causing iron deposition in those regions are unknown. Possibly, the aging condition can synthesize some special type of neurons to increase the expression of iron-storing protein ferritin. For example, globus pallidus has rich iron level as equivalent as in the hepatocytes. Different scavenger cells including macrophages can become iron-rich when the brain cells started to undergo apoptosis. IRP1 and IRP2 may also play important roles in controlling the cellular iron homeostasis in the brain. Alterations in these IRPs signaling pathways can contribute to higher iron deposition in the brain. For example, iron deficiency condition can allow IRP1 to interact IREs for up-regulating the TfR expression in the brain endothelial cells, but alterations in these signaling pathways can increase the brain iron accumulation.

CSF is the major route for iron export from the brain into the blood from the subarachnoid space. Interference in the CSF route can cause the over iron accumulation in the brain. For example, transferrin is very low in the CSF, thus exporting iron from the brain to blood is limited. Some other iron-storing proteins like lactoferrin, ferritin, and non-protein-bound iron can help to export the iron, but these proteins function is majorly affected by some pathological conditions. In addition, microglia and other phagocytic cells are involved to export the iron. Mitochondrial proteins can also involve in exporting iron. For example, iron/sulfur (Fe/S) clusters are acted as a prosthetic group of respiratory complexes which can contribute to eliminate the iron from mitochondria. In other way, the storage protein ferritin can reduce the free from iron in the cytosol by sequestering them, but changes in this export mechanism can contribute to increase iron accumulation in the brain. In a simple description, age factors, genetic mutation, and inflammation can cause the iron accumulation in the brain [56, 62].

1.4 Iron Metabolism and Brain Diseases

Iron deposition in a particular region of the brain can mishandle the iron metabolism because this region is already rich with iron, and thereby causing various distinct disorders called neurodegeneration with brain iron accumulation (NBIA). Neurodegeneration with brain iron accumulation is mainly linked with the mutations in the iron homeostasis genes and is characterized by gait problems, movement dysfunction, and motor–cognitive problems. Until now, there are fifteen major genetically distinct disorders which can be discussed in further content. However, other neurodegenerative diseases like Parkinson's disease, Alzheimer's disease, and Hunting's disease can also be believed to be caused by secondary iron accumulation. This is not necessarily the case. It may also be a disease caused by existing iron deposits (such as age and inflammation).

1.4.1 Iron Metabolism and Parkinson's Disease

Parkinson's disease (PD) is one of the most extensive diseases that relate to dysregulation of iron metabolism. Although the causes of PD have been very clear, the initial discovery of PD was written in 1817 by Parkinson [48]. After this, Jean-Martin Charcot expanded this research internationally [5]. Since then, several concepts are associated with PD causes. Recent research has received immense interest to find the role of iron in causing PD. Patients with PD have higher iron accumulation in the pars compacta area of the substantia nigra. Several MRI studies have also observed that elevated level of iron is associated with PD. The cause of PD is linked with the iron-binding ability of the neuromelanin, a dark pigment present in the dopaminergic neurons, and can sequestrate the reactive irons in the cytosol of neurons. Reduced binding capacity of neuromelanin to iron can increase the availability of free iron in the substantia nigra and consequent oxidative damages and mitochondrial dysfunction, and thus neuronal cell death. Friedman et al. suggested that despite increasing ferritin-bound iron, free reactive iron can also contribute to set the PD. It has been observed that Lewy bodies present in the substantia nigra of PD condition, and the formation of Lewy bodies may be occured with the iron interaction to ubiquitinated alpha-synuclein. In the earlier stages of PD, alpha-synuclein can be protective, but the interaction of iron with alpha-synuclein can increase the oxidative damage and protein aggregation. Studies found that iron can stimulate alpha-synuclein in the post-mortem brain. Furthermore, iron chelation can partially reverse the alpha-synuclein aggregation. However, some studies have observed that there is no iron accumulation in the earlier stages of PD, suggesting a puzzling role of iron in the PD. In addition, loss of neuromelanin binding to iron can cause oxidative stress through Fenton reaction, resulting in neurodegeneration.

1.4.2 Role of Iron in Alzheimer's Disease

Abnormal iron metabolism has been considered as a vital marker for the earlier setting of Alzheimer's disease (AD). However, the link of iron accumulation to the AD pathology is not well established. Studies have observed that the patients with AD have elevated level of iron in the cortical, subcortical, and white matters of the brain. Amyloid (Aβ) plaque aggregation in a specific region of the brain has been considered as AD pathology, and Aβ is formed by proteolytic cleavage of amyloid precursor protein (APP). APP is cleaved by α-secretase and β-secretase. Normally, alpha pathway is not involved in the generation of amyloidogenesis, whereas beta pathway is involved in the amyloidogenesis. Overproduction of Aβ42 and alloforms of Aβ can lead the AD pathogenesis, and the connection of APP and Aβ alloforms with abnormal iron metabolism has been observed in AD patients. Both ferrous and ferric forms of iron can interact with the Aβ alloforms. However, Aβ42 fragments can form amyloid more rapidly. Aggregation of Aβ42 can form monomer, dimers, oligomers, fibrils, and senile fibril plaques. The ferric form of iron can bind with all forms of Aβ42. However, it has a difficulty to break from the fibrils, but this plaque formation with ferric form can initially remove the excess-free form of iron in the brain. However, the ferrous form can also bind with Aβ42 to revert the function of amyloid formations. The core of matured amyloid plaque has both ferrous and ferric form of iron, and the core of amyloid plaque is the place of oxidative stress. In addition, the ferroxidase activity of APP can prevent the FPN1-mediated iron loading to Tf resulting in iron retention in the neurons and further causing neuronal death. Alzheimer's disease brains have increased the accumulation of iron in the specific region of the brain such as the hippocampus and cerebral cortex. Aβ plaques and neurofibrillary tangles have reactive iron and are the important site for catalytic reactivity. In addition, iron-binding proteins such as transferrin, ferritin, and IRP2 are linked with Alzheimer's neurodegeneration. Studies have observed that ferritin-containing microglia present in the Alzheimer brains, thus suggesting a role of ferritin in the disruption of the iron homeostasis in the Alzheimer brains.

1.4.3 Iron Overloading in Amyotrophic Lateral Sclerosis

Amyotrophic lateral sclerosis (ALS), a neurodegenerative disease, is characterized by gradual breakdown of nerve cells and cell death. Patients with ALS have higher iron accumulation in the CNS, and several iron regulatory molecules level is altered with ALS condition [31, 36]. Dysregulation of iron metabolism can activate microglia and oxidative stress. At the point, when microglia catabolize the cell debris, it can express a higher level of ferritin in the ALS. A study observed that iron overloaded in the activated microglia of motor cortex area, particularly upper motor neuron cell bodies.

1.4.4 Gene Mutation Induced Disease

Genetic analysis of iron overload condition has revealed a number of target proteins that control iron homeostasis, and the mutation in these proteins disrupts the iron homeostasis. Molecular level of research has identified the several loops that control the iron homeostasis and is mounted in the hepcidin–ferroportin interaction. This regulatory pathway accounts genetic mutation of iron overloading condition and related disorder. PKAN is a neuronal iron overloading disease linked with the gene encoding pantothenate kinase 2 (PANK2) mutations, a crucial enzyme present in the mitochondrial inter-membrane space and synthesis mitochondrial coenzyme A from pantothenate. PKAN is previously known as Hallervorden–Spatz syndrome. This disease is characterized by neurological impairments such as childhood onset dystonia and spasticity. Study has demonstrated that lack of Pank2$^{-/-}$ in the mice did not have iron overload causing neurological symptoms in the CNS [64]. Patients with PKAN have reduced cholesterol level and fatty acid synthesis, and also the Pank2$^{-/-}$ mice have reduced mitochondrial potential and ATP generation. However, the exact reason of iron overload in the brain regions is not entrenched, particularly, why PANK2 associated iron overload in the globus pallidus, the possible reason is that the other homologues such as PANK1 and PANK3 may be too low to compensate the loss of PANK2. Likewise, the neurons can cause the iron overload before they die, which results in iron overload neuron loss in PKAN.

PLAN is linked with different mutations in the PLA2G6 gene, and the symptoms are dystonia, spasticity, and cerebellar atrophy. PLA2G6 gene encodes phospholipase A2 group VI, a critical regulator of mitochondrial membrane stability and remodeling. MRI result has shown that iron deposition presents in the globus pallidus and substantia nigra and striatum varies. PLA2G6 gene producing enzymes regulate fatty acid synthesis in the brain. However, it is believed that acetyl CoA can recompense the needs of fatty acid in the brain. MPAN is a condition associated with childhood or early adulthood. MPAN affected people can have movement problems including spasticity and dystonia. Other neurological issues such as degeneration in the nerves that carry information to the brain (optic atrophy), dysphagia and incontinence are linked with MPAN. Mutations in the C19orf12 can cause MPAN. C19orf12 encodes a mitochondrial membrane protein which can participate in the fatty acid metabolism and branched-chain amino acids. MPAN is characterized by iron over accumulation in the globus pallidus and substantia nigra. In MPAN, Lewy bodies containing abnormal alpha-synuclein present throughout in the globus pallidus, deep gray matter, and neocortex.

Aceruloplasminemia is characterized by the mutation of CP. It was initially described in 1987. Ceruloplasmin is a multicopper ferroxidase, lost its ferroxidase activity in aceruloplasminemia condition. As a result, CP may not be able to convert the ferrous iron form into ferric form to transferrin, which results to fail the export iron out of brain astrocytes and macrophages to the circulatory system. Patients with aceruloplasminemia can have higher iron overload in the astrocytes, and these regions of the brain lose the neurons, which result in the triad of adult-onset neurological disease and retinal degeneration. The deformity of astrocytes is a prominent characteristic of aceruloplasminemia brains. The proportion of iron deposition is correlated with the globular structure of the astrocytes [41]. In addition, the iron does not reach to the neurons due to inactivity of CP, and thus, neurons are susceptible to die.

Neuroferritinopathy is an autosomal dominant neurodegenerative disease caused by ferritin light chain 1 (FTL1) light chain mutations. Curtis and groups identified first time neuroferritinopathy in 2001 [7]. Until now, there have been ten mutations reported to cause this disease. Additional nucleotide inclusion due to a mutation in the FTL1 can neglect to sequester the iron, and consequently iron leaks out of the ferritin. This condition makes the cytosol to turn out to be progressively oxidized environment. Ferritin can also respond to this oxidized cytosolic environment, and this can further allow more leakage of iron. Furthermore, higher accumulation of iron can make the IRE-binding proteins to not turn IRE-binding form which further results in the overproduction of ferritin and ferritin aggregation with iron in the neurons, substantia nigra, glial cells of the globus pallidus, and cerebellum. Initial failure of iron sequestration can lead the higher transcription, and reduced translational repression in the FTL can allow this abnormal protein segregation in the various regions of the brain. FAHN is caused by the mutation in the gene encoding fatty acid-2-hydroxylase gene (FA2H) and is involved to produce fatty acid for sphingolipids. Defect in this gene is associated with abnormal myelination. Studies have shown that FAHN patients have iron accumulation in the globus pallidus, substantia nigra, and subcortical and periventricular regions. Also, ferritin-loaded microglial cells enter into the areas of regenerating tissue for further iron loading. FAHN is characterized by the spasticity, ataxia/dystonia, and optic atrophy.

1.5 Current Bottlenecks in the Prevention and Treatment of Neurological Diseases

1.5.1 The Evaluation of Neurological Therapies

Several serendipitous findings are observed for treating neurological diseases because these conditions are different from other disorders, and the evaluation therapy has been designed based on the developing information of biochemistry, physiology, and anatomy of the brain. For example, Parkinson's disease is one of the important neurological conditions, and treatment for this disease is tightly based on the above-mentioned merit. Parkinson was believed that the "progress of the disease may be stopped" and he deliberately inserted a small piece of cork into the blisters and inflammatory skin for expelling the pressure away from the brain and spinal cord. Charcot and groups were first established the treatment for PD using belladonna alkaloids. These alkaloids influence the cholinergic/dopaminergic balance in the striatum for improving PD. Since then, many active anticholinergic agents have been developed for treating PD. However, Charcot wanted to use hyoscyamine. He occasionally used hyoscyamine with rye-based ergot products, a pharmacological dopamine agonist used to invigorate the striatal dopamine receptors. He inferred in 1872 that he attempted almost everything to treat PD, but all of them were rejected. Then, Charcot tried vibratory therapy for managing PD, and he found that patients with PD had decreased symptoms. However, the tremor was not improved. He used a combination of hyoscyamine with arsenic, morphia and *cannabis,* and opium and he observed that the patients with PD have distinct improvements [19]. The modern therapy is used cannabis for dopaminergic activation. Although several anticholinergic drugs were developed in the nineteenth and twentieth centuries, all of them were similar effects. Holtz discovered an important enzyme called dopa carboxylase that can convert levodopa to dopamine [29]. Loss of dopamine leads the pathogenesis of PD. Ehringer and Horneykiewicz observed that the depletion of dopamine in the human brains [13, 28]. Birkmayer and Horneykiewicz injected the levodopa intravenously first time to the PD patients, and they observed significant improvement in the PD [3]. For searching levodopa in the naturally available plants, Manyam observed that cowitch plant (*Mucuna pruriens*) contains levodopa [37]. It is pointed out that the current treatment of PD defects, especially the late use of L-DOPA, has no effect, and the understanding of iron metabolism is helpful for the treatment of PD.

1.5.2 Alzheimer's Disease

AD is one of the most leading dementia diseases. Although the number of scientific effort has been made to prevent the disease, there are no effective pharmacotherapeutic options available for treating AD. To date, several drugs have been prescribed based on symptomatic approach for counterbalance the neurotransmitter disturbances in the AD. Cholinergic hypothesis has been linked with AD including loss of acetylcholine neurons and acetylcholine synthesis and degradation. There are three approved drugs named cholinesterase inhibitors (CI) used for enhancing the cholinergic transmission like donepezil, rivastigmine, and galantamine. These CIs have been shown to improve the cognitive functions of AD. Another therapeutic option for AD is memantine, and this drug can bind with N-methyl-D-aspartate antagonist for protecting neurons. In addition, a combination of donepezil with memantine significantly improved the cognitive function of moderate to severe AD patients [27, 54]. Another cause for AD is amyloid (Aβ) plaque aggregation. Several researches have been driven to finding a compound for preventing Aβ aggregation. The only Aβ aggregation inhibitor is a synthetic glycosaminoglycan 3-amino-1-propanesulfonic acid (3APS, tramiprosate). Tramiprosate interferes with the Aβ aggregation. However, data suggest that tramiprosate increases the abnormal aggregation of Aβ [53]. Other compounds such as Colostrinin and scyllo-Inositol can interfere with the Aβ aggregation. However, the specificity of those compounds is questionable. Drugs such as PBT2 is a second-generation drug of 8-OH quinoline interferes with the Cu^{2+} and Zn^{2+} mediated toxic oligomerization of Aβ. Also, several compounds that inhibit BACE (β-site AP-cleaving enzyme inhibition) and γ-secretase are in the preclinical stage. Therefore, the current clinical trials for the treatment of AD drugs have failed because the mechanism of AD pathogenesis has not been fully understood, so the understanding of iron metabolism may contribute to the development of AD drugs.

1.5.3 NBIA Therapies

Therapies for NBIA disorders are symptomatic. NBIA is associated with different genes mutations that are linked with iron metabolism-regulating proteins. Therefore, understanding a common feature that shares to induce the pathogenesis in NBIA can help to develop therapies for NBIA disorders. For example, aceruloplasminemia and neuroferritinopathy have common defects in handling iron, which results to increase the oxidative damages, whereas other NBIA disorders such as PLAN and MPAN are linked with lipid metabolism and mitochondria. Therefore, finding a common link with membrane homeostasis, mitochondrial function, and iron metabolism may advantage therapeutic development. Oxidative stress is one of the common conditions of above-mentioned situations. Therefore, understanding of the mechanism that induces iron metabolism causing oxidative stress and its further consequences in developing NBIA may help to design a therapy for NBIA-associated disorders.

1.6 Conclusion

The present drugs for treating iron overload causing neurodegeneration are symptomatic and do not able to prevent the progression of the diseases. Nevertheless, these drugs are giving consistent improvements for cognition and global status. In addition, the search for disease-modifying approach for treating these diseases is unsuccessful in demonstrating the efficacy in the clinical stages. Meanwhile, the research is focusing to target other possible mechanisms that involve progress iron causing neurodegeneration such as iron overloading neuroinflammation and oxidative stress. For example, clinical trials of antioxidants usage such as Vitamin E and ω3 fatty acids did not show any significant improvements in patients with neurodegenerative diseases. Before developing a novel compound to treat iron causing neurodegeneration, it must be needed profound investigations of underlying mechanisms of pathogenesis of those neurological diseases. For example, disease-modifying approach has been failed in stage III of the clinical trial for proving its efficacy against AD pathogenesis. It needs a better understanding of tau, Aβ, and other direct and indirect factors to develop successful disease-modifying drugs. Altogether, new strategies need to be focused to examine the neuroprotective activity of disease-modifying agents in the presymptomatic stages of neurological diseases. In the development of drugs for the treatment of these diseases, it is proposed to consider the joint use of drugs and methods for regulating iron metabolism to improve the prevention and treatment of diseases.

References

1. Andersen HH, Johnsen KB, Moos T (2014) Iron deposits in the chronically inflamed central nervous system and contributes to neurodegeneration. Cell Mol Life Sci 71:1607–1622
2. Appleton TC, Morgan E, Baker E (1971) A morphological study of transferrin uptake by reticulocytes. In: Trávníèek T, Neuwirt J (eds) The regulation of erythropoiesis and haemoglobin synthesis. Universita karlova, prague, pp 310
3. Birkmayer W, Hornykiewicz O (1961) The L-3,4-dioxyphenylalanine (DOPA)-effect in Parkinson-akinesia. Wien Klin Wochenschr 73:787–788
4. Busacchi V (1958) Vincenzo Menghini and the discovery of iron in the blood. Bull Sci Med (bologna) 130:202–205
5. Charcot JM (1872) De la paralysie agitante. In: Oeuvres Complètes (t1) Leçons sur les maladies du système nerveux, pp 155–188. Delahaye A, Paris: In: English: Charcot JM. 1877. On Parkinson's disease. In Lectures on diseases of the nervous system delivered at the Salpêtrière (trans: Sigerson G). New Sydenham Society, London, pp 129–156
6. Chen H, Attieh ZK, Syed BA, Kuo YM, Stevens V, Fuqua BK, Andersen HS, Naylor CE, Evans RW, Gambling L et al (2010) Identification of zyklopen, a new member of the vertebrate multicopper ferroxidase family, and characterization in rodents and human cells. J Nutr 140:1728–1735
7. Curtis AR, Fey C, Morris CM, Bindoff LA, Ince PG, Chinnery PF, Coulthard A, Jackson MJ, Jackson AP, McHale DP et al (2001) Mutation in the gene encoding ferritin light polypeptide causes dominant adult-onset basal ganglia disease. Nat Genet 28:350–354

8. Conrad ME Jr, Crosby WH (1963) Intestinal mucosal mechanisms controlling iron absorption. Blood 22:406–415
9. Conrad S, Genth H, Hofmann F, Just I, Skutella T (2007) Neogenin-RGMa signaling at the growth cone is bone morphogenetic protein-independent and involves RHoA, ROCKk, and PKC. J Biol Chem 282:16423–16433
10. Deane R, Zheng W, Zlokovic BV (2004) Brain capillary endothelium and choroid plexus epithelium regulate transport of transferrin-bound and free iron into the rat brain. J Neurochem 88:813–820
11. Donovan A, Brownlie A, Zhou Y, Shepard J, Pratt SJ, Moynihan J, Paw BH, Drejer A, Barut B, Zapata A, Law TC, Brugnara C, Lux SE, Pinkus GS, Pinkus JL, Kingsley PD, Palis J, Fleming MD, Andrews NC, Zon LI (2000) Positional cloning of zebrafish ferroportin1 identifies a conserved vertebrate iron exporter. Nature 403:776–781
12. Drysdale JW, Munro HN (1965) Failure of actinomycin D to prevent induction of liver apoferritin after iron administration. Biochim Biophys Acta 103:185–188
13. Ehringer H, Hornykiewicz O (1960) Verteilung von noradrenalin und dopamin im gehirn des menschen und ihr verhalten bei erkrankungen des extrapyramidalen systems. Klin Wschr 38:1126–1239
14. Fenton HJH (1876) On a new reaction or tartaric acid. Chemical News 33:190
15. Fenton HJH (1893) The oxidation of tartaric acid in presence of iron. Chem Soc Proc 9:113
16. Fineberg RA, Greenberg DM (1955) Ferritin biosynthesis II. Acceleration of synthesis by the administration of iron. J Biol Chem 214:97–106
17. Finch CA (1959) Body iron exchange in man. J Clin Invest 38:392–396
18. Fontés G, Thivolle I (1925) Sur la teneur du sérum en fer non hémoglobinique et sur sa diminution au cours de l'anémie expérimentale. Compt Rend Soc Biol 96:687
19. Gowers WR (1899) Paralysis agitans. In: Allbutt A, Rolleston T (eds) A system of medicine. Macmillan, London, pp 156–178
20. Guo B, Phillips JD, Yu Y, Leibold EA (1995) Iron regulates the intracellular degradation of iron regulatory protein 2 by the proteasome. J Biol Chem 270:21645–21651
21. Hahn PF, Bale WF, Ross JF, Balfour WM, Whipple GH (1943) Radioactive iron absorption by gastrointestinal tract: influence of anemia, anoxia, and antecedent feeding distribution in growing dogs. J Exp Med 78:169–188
22. Hahn P, Qian Y, Dentchev T, Chen L, Beard J, Harris ZL et al (2004) Disruption of ceruloplasmin and hephaestin in mice causes retinal iron overload and retinal degeneration with features of age-related macular degeneration. Proc Natl Acad Sci U S A 101:13850–13855
23. Halliwell B, Gutteridge JM (1984) Role of iron in oxygen radical reactions. Methods Enzymol 105:47–56
24. Hentz MW, Kuhn LC (1996) Molecular control of vertebrate iron metabolism: mRNA-based regulatory circuits operated by iron, nitric oxide, and oxidative stress. Proc Natl Acad Sci U S A 93:8175–8182
25. Hentze MW, Muckenthaler MU, Galy B, Camaschella C (2010) Two to tango: regulation of mammalian iron metabolism. Cell 142:24–38
26. Holmberg CG, Laurell CB (1945) Studies on the capacity of serum to bind iron—a contribution to our knowledge of the regulation mechanism of serum iron. Acta Physiol Scand 10:307
27. Howard R, McShane R, Lindesay J, Ritchie C, Baldwin A, Barber R, Burns A, Dening T, Findlay D, Holmes C, Hughes A, Jacoby R, Jones R, Jones R, McKeith I, Macharouthu A, O'Brien J, Passmore P, Sheehan B, Juszczak E, Katona C, Hills R, Knapp M, Ballard C, Brown R, Banerjee S, Onions C, Griffin M, Adams J, Gray R, Johnson T, Bentham P, Phillips P (2012) Donepezil and meantime for moderate-to-severe Alzheimer's disease. N Engl J Med 366:893–903
28. Hornykiewicz O (2002) Dopamine miracle: from brain homogenate to dopamine replacement. Mov Disord 17:501–508
29. Holtz P, Heise R, Lüdtke K (1938) Fermentative degradation of l-dioxyphenylalanine (dopa) by kidney. Naunyn-Schmiedeberg's Arch. Pharmacol 191:87–118
30. Jandl JH, Katz JH (1963) The plasma-to-cell cycle of transferrin. J Clin Invest 42:314–326

31. Kasarskis EJ, Tandon L, Lovell MA, Ehmann WD (1995) Aluminum, calcium, and iron in the spinal cord of patients with sporadic amyotrophic lateral sclerosis using laser microprobe mass spectroscopy: a preliminary study. J Neurol Sci 130:203–208
32. Ke Y, Qian ZM (2007) Brain iron metabolism: neurobiology and neurochemistry. Prog Neurobiol 83:149–173
33. Leibold EA, Munro HN (1988) Cytoplasmic protein binds in vitro to a highly conserved sequence in the 5′ untranslated region of ferritin heavy- and light-subunit mRNAs. Proc Natl Acad Sci U S A. 85:2171–2175
34. Lecanu LR, l'hématosine de (1830) O matière colorante du sang. Ann Chim 45:5–27
35. Lecanu LR (1831) Nouvelle recherches sur le sang. Ann Chim 48:308–327
36. Li X, Jankovic J, Le W (2011) Iron chelation and neuroprotection in NDs. J Neural Trans 118:473–477
37. Manyam BV (1990) Paralysis agitans and levodopa in "Ayurveda": ancient Indian medical treatise. Mov Disord 5(1):47–48
38. Malecki EA, Devenyi AG, Beard JL, Connor JR (1999) Existing and emerging mechanisms for transport of iron and manganese to the brain. J Neurosci Res 56:113–122
39. McKie AT, Marciani P, Rolfs A, Brennan K, Wehr K, Barrow D, Miret S, Bomford A, Peters TJ, Farzaneh F, Hediger MA, Hentze MW, Simpson RJ (2000) A novel duodenal iron-regulated transporter, IREG1, implicated in the basolateral transfer of iron to the circulation. Mol Cell 5:299–309
40. McKie AT, Barrow D, Latunde-Dada GO, Rolfs A, Sager G, Mudaly E, Mudaly M, Richardson C, BarlowD Bomford A (2001) An iron-regulated ferric reductase associated with the absorption of dietary iron. Science 291:1755–1759
41. Mayajima H (2003) Aceruloplasminemia an iron metabolic disorder. Neuropathology 23:345–350
42. Monarde N (1925) Joyful newes out of the newe founde worlde. New York
43. Morgan EH, Appleton TC (1969) Autoradiographic localization of 125-I-labelled transferrin in rabbit reticulocytes. Nature 223:1371–1372
44. Mccance RA, Widdowson EM (1938) The absorption and excretion of iron following oral and intravenous administration. J Physiol 94:148–154
45. Nicolas G, Bennoun M, Devaux I et al (2001) Lack of hepcidin gene expression and severe tissue iron overload in upstream stimulatory factor 2 (USF2) knockout mice. Proc Natl Acad Sci U S A. 98(15):8780–8785
46. Oshiro S, Kawahara M, Kuroda Y, Zhang C, Cai Y, Kitajima S, Shirao M (2000) Glial cells contribute more to iron and aluminum accumulation but are more resistant to oxidative stress than neuronal cells. Biochim Biophys Acta 1502:405–414
47. Owen D, Kühn LC (1987) Noncoding 3′ sequences of the transfer in receptor gene are required for mRNA regulation by iron. EMBO J 6:1287–1293
48. Parkinson J (1817) An essay on the shaking palsy. Whittingham and Rowland for Sherwood, Needly and Jones, London
49. Pollycove M, Mortimer R (1961) The quantitative determination of iron kinetics and hemoglobin synthesis in human subjects. J Clin Invest 40:753–782
50. Qian ZM, Chang YZ, Zhu L, Yang L, Du JR, Ho KP et al (2007) Development and iron-dependent expression of hephaestin in different brain regions of rats. J Cell Biochem 102:1225–1233
51. Rouault TA (2006) The role of iron regulatory proteins in mammalian iron homeostasis and disease. Nat Chem Biol 2:406–414
52. Schade AL, Caroline I (1944) Raw hen egg white and the role of iron in growth inhibition of *Shigella dysenteriae*, *Staphylococcus aureus*, *Escherichia coli* and *Saccharomyces cerevisiae*. Science 100:14–15
53. Santa-Maria I, Hernández F, Del Rio J, Moreno FJ, Avila J (2007) Tramiprosate, a drug of potential interest for the treatment of Alzheimer's disease, promotes an abnormal aggregation of tau. Mol Neurodegener 6:17

54. Tariot PN, Farlow MR, Grossberg GT, Graham SM, McDonald S, Gergel I (2004) Memantine treatment in patients with moderate to severe Alzheimer disease already receiving donepezil: a randomized controlled trial Memantine Study Group. JAMA 291(3):317–324
55. Troadec MB, Warner D, Wallace J et al (2011) Targeted deletion of the mouse Mitoferrin1 gene: from anemia to protoporphyria. Blood 117(20):5494–5502
56. Vanotti A, Delachaux A (1949) iron metabolism and its clinical significance. Grune & Stratton, New York
57. Vivot RM, Goitia B, Usach V, Setton-Avruj PC (2013) DMT1 as a candidate for non-transferrin-bound iron uptake in the peripheral nervous system. BioFactors 39:476–484
58. Vulpe CD, Kuo YM, Murphy TL, CowleyL Askwith C, Libina N, Gitschier J, Anderson GJ (1999) Hephaestin, a ceruloplasmin homologue implicated in intestinal iron transport, is defective in the sla mouse. Nat Genet 21:195–199
59. Wächtershäuser G (1990) Evolution of the first metabolic cycles. Proc Natl Acad Sci U S A 87:200–204
60. Wächtershäuser G (2000) Origin of life. Life as we don't know it. Science 289:1307–1308
61. Wang SM, Fu LJ, Duan XL, Crooks DR, Yu P, Qian ZM, Di XJ, Li J, Rouault TA, Chang YZ (2010) Role of hepcidin in murine brain iron metabolism. Cell Mol Life Sci 67(1):123–133
62. You LH, Yan CZ, Zheng BJ, Ci YZ, Chang SY, Yu P, Gao GF, Li HY, Dong TY, Chang YZ (2017) Astrocyte hepcidin is a key factor in LPS-induced neuronal apoptosis. Cell Death Dis 8(3):e2676
63. Zahringer J, Baliga BS, Munro HN (1976) Novel mechanism for translational control in regulation of ferritin synthesis by iron. Proc Natl Acad Sci USA 73:857–861
64. Zorzi G, Zibordi F, Chiapparini L, Bertini E, Russo L, Piga A et al (2011) Iron-related MRI images in patients with pantothenate kinase-associated neurodegeneration (PKAN) treated with deferiprone: results of a phase II pilot trial. Mov Disord 26:1756–1759

Chapter 2
Cellular Iron Metabolism and Regulation

Guofen Gao, Jie Li, Yating Zhang and Yan-Zhong Chang

Abstract Iron is an essential trace element in the human body, but excess iron is toxic as it contributes to oxidative damage. To keep iron concentration within the optimal physiologic range, iron metabolism at the cellular level and the whole systemic level are tightly regulated. Balance of iron homeostasis depends on the expression levels and activities of iron carriers, iron transporters, and iron regulatory and storage proteins. Divalent metal transporter 1 (DMT1) at the apical membrane of intestinal enterocyte brings in non-heme iron from the diet, whereas ferroportin 1 (FPN1) at the basal membrane exports iron into the circulation. Plasma transferrin (Tf) then carries iron to various tissues and cells. After binding to transferrin receptor 1 (TfR1), the complex is endocytosed into the cell, where iron enters the cytoplasm via DMT1 on the endosomal membrane. Free iron is either utilized in metabolic processes, such as synthesis of hemoglobin and Fe–S cluster, or sequestered in the cytosolic ferritin, serving as a cellular iron store. Excess iron can be exported from the cell via FPN1. The liver-derived peptide hepcidin plays a major regulatory role in controlling FPN1 level in the enterocyte, and thus controls the whole-body iron absorption. Inside the cells, iron regulatory proteins (IRPs) modulate the expressions of DMT1, TfR1, ferritin, and FPN1 via binding to the iron-responsive element (IRE) in their mRNAs. Both the release of hepcidin and the IRP–IRE interaction are coordinated with the fluctuation of the cellular iron level. Therefore, an adequate and steady iron supplement is warranted for the utilization of cells around the body. Investigations on the molecular mechanisms of cellular iron metabolism and regulation could advance the fields of iron physiology and pathophysiology.

Keywords Iron · DMT1 · TfR1 · FPN1 · Ferritin · IRPs · Hepcidin

G. Gao (✉) · J. Li · Y. Zhang · Y.-Z. Chang (✉)
Laboratory of Molecular Iron Metabolism, College of Life Sciences, Hebei Normal University, Shijiazhuang, Hebei Province, China
e-mail: guofen83@hotmail.com

Y.-Z. Chang
e-mail: chang7676@163.com

Y.-Z. Chang (ed.), *Brain Iron Metabolism and CNS Diseases*,
Advances in Experimental Medicine and Biology 1173,
https://doi.org/10.1007/978-981-13-9589-5_2

2.1 Body Iron and Its Metabolism

Iron (Fe) is one of the essential trace metals in the human body, which is found in the active centers of many enzymes and oxygen carrier proteins [1]. Iron is a critical component of cytochromes a, b, and c, cytochrome oxidase, and iron–sulfur (Fe–S) complexes of the oxidative chain, therefore important for adenosine triphosphate (ATP) production. Iron is a co-enzyme in ribonucleoside reductase which commits to DNA synthesis. Iron is also involved in TCA cycle since the succinate dehydrogenase and aconitase are both Fe-dependent enzymes. In addition, the iron in the brain is required for myelogenesis and myelin maintenance as well. Therefore, a constant and readily available supply of iron is important for all the organs of the body [2]. The iron transport, storage, and regulation are tightly regulated at the cellular and systemic levels [1, 2].

Human bodies contain 4–5 g of iron on average. Dietary iron is absorbed predominately in the duodenum and enters blood circulation in the small intestine. The absorption of heme iron is poorly understood. Non-heme iron is transported across the apical membrane of the intestinal enterocyte by divalent metal transporter 1 (DMT1) and is exported into the circulation via ferroportin 1 (FPN1). Once in blood circulation, iron binds to apotransferrin and forms Fe-transferrin (Tf) complex [3]. Serum Tf is the major vehicle for iron transport in the body and carries iron to other cells and tissues through the circulation [3]. At the target cell, Tf binds to transferrin receptor 1 (TfR1) on the cell membrane, and the TfR1-Tf-Fe complex is then endocytosed into the cell, where the iron is released [4]. Free iron either enters mitochondrion for utilization in metabolic processes, such as synthesis of hemoglobin and Fe–S cluster, or is incorporated into the cytosolic iron storage protein, ferritin and serves as a cellular store of iron [5]. Excess iron is transported out of the cell by iron efflux protein ferroportin 1 (FPN1) located on the cell membrane [6].

About 2.5 g iron in the body is contained in the hemoglobin, the red pigment in red blood cells, which is needed to carry oxygen through the blood. This portion of iron is circulated in the body by macrophages after they engulf the senile erythrocytes and release iron. A relatively small amount iron (3–4 mg) circulates through the plasma, bound to Tf. Most of the rest iron is stored in ferritin complexes that are present in all cells, but most common in bone marrow, liver, and spleen [1, 5]. Of these iron stores, the iron in the liver is the primary physiologic source of reserve iron in the body. There are a lower non-hemoglobin hemoproteins, such as myoglobin, cytochromes, catalases, heme peroxidase, and endothelial nitric oxide synthase, involved in diverse biological functions including the transportation of diatomic gases, chemical catalysis, diatomic gas detection, and electron transfer [7]. Because of its toxicity, free soluble iron in the body is kept in low concentration in the body.

2.2 Regulation of Iron Metabolism

Iron homeostasis in the body is regulated at two different levels, the systemic level and the cellular level [8]. The systemic iron level is mainly balanced by the controlled absorption of dietary iron by intestinal enterocytes, the cells that line at the interior of the intestines. The cellular iron level is controlled differently by different cell types due to the expression of particular iron regulatory and transport proteins [9].

2.2.1 Systemic Iron Regulation

At the systemic level, iron homeostasis involves the regulation of a peptide 'hormone' hepcidin [10]. Hepcidin is mainly produced by hepatocytes in response to high serum iron concentration, inflammatory stimuli, or hypoxia [10, 11]. It binds to the extracellular loop of FPN1 and causes its internalization and degradation, and thereby reduces cellular iron efflux [2, 12]. Hepcidin regulates iron homeostasis mainly via acting on FPN1 of intestinal enterocytes and FPN1 of macrophages, thereby affecting iron intake from the intestine and iron release from macrophages [13, 14]. The level of hepcidin is also regulated by the systemic iron level [15]. When body iron is overloaded, the level of hepcidin increases to reduce circulating iron. When body iron is deficient, the level of hepcidin decreases to promote small intestinal iron intake. In addition to the iron level, inflammation, hypoxia, and erythropoiesis regulate the expression of hepcidin as well [16].

2.2.2 Cellular Iron Regulation

Cellular iron homeostasis is tightly regulated by iron regulatory proteins (IRPs), including IRP1 and IRP2, which subsequently regulate the levels of iron transporters, TfR1, divalent metal transporter 1 (DMT1) and FPN1, and ferritin as well [17]. Generally, when the cellular iron concentration is low, the active center of IRPs binds to the stem-loop structure of the iron-responsive element (IRE) located at the 3′-untranslated region (UTR) of TfR1 and DMT1 mRNAs. This binding stabilizes TfR1 and DMT1 mRNAs and increases its cellular expression, thereby increasing iron uptake [18]. When the iron concentration is high, the active center of IRPs is occupied by four Fe–S, which blocks the binding of IRPs to IRE of TfR1 or DMT1, resulting in low TfR1 and DMT1 translation levels, and thus reduces iron uptake [18]. The IRP/IRE system also regulates the stability of FPN1 and ferritin. However, the binding of IRPs to the IRE of FPN1 or ferritin mRNA located at their 5′-UTR suppresses their translation and causes lower protein expression [18]. Thus, IRPs play a key role in the maintenance of cellular iron homeostasis. Studies have shown that the IRPs knockout mice had significant misregulation of iron metabolism and developed neurodegeneration [18, 19].

In the intestine, the control of ferritin and FPN1 expression by IRPs limits intestinal iron absorption and thus controls the whole-body iron level [20]. In the bone marrow, IRPs regulate the translation of erythroid 5-aminolevulinate synthase's mRNA; therefore, the erythroid heme biosynthesis is coordinated with iron availability. Besides, the translational control of HIF2α mRNA in the kidney by IRP1 coordinates erythropoietin synthesis with iron and oxygen supply [19, 20].

2.3 Iron Metabolism-Related Proteins

2.3.1 Iron Transport Protein-Tf

Transferrin (Tf) is a serum glycoprotein with a molecular weight of 80 kDa, which is mainly produced by the liver. It is an important iron chelating protein with serum iron transport function. Tf transports iron in the blood by picking up free ferric iron (Fe^{3+}) and delivering it into cells [3]. Tf can reversibly bind to Fe^{3+}. Once binding, a conformation change in Tf will occur, which facilitates its recognition and binding to its receptor-TfR1 on the cell membrane [3, 4]. The binding of Tf to TfR1 will initiate a receptor-mediated endocytotic process, in which the Tf-TfR1 complex is internalized. Iron is released in the endosome, and the complex is recycled to the cell surface where the Tf is released [4, 9]. Thus, iron is transported to various organs and tissues of organisms through blood transportation and utilized for various life activities.

2.3.2 Iron Uptake Proteins-TfR1 and DMT1

Transferrin receptor (TfR1) is a transmembrane single-chain glycoprotein with two disulfide subunits, which is composed of ~700 amino acids and has a molecular weight of about 95 kDa, responsible for iron transport in cells [21, 22]. TfR1 specifically recognizes and binds with Tf at its extracellular domain, with the highest affinity to Tf carried two Fe^{3+} ions [4]. The complex formed by TfR1 and Tf initiates the formation of endocytosis vesicles on the membrane and finishes the endocytosis process. After iron dissociation, reduction, and translocation, ferrous iron (Fe^{2+}) was released via DMT1, and then TfR1 was recycled to the surface of the cell membrane. There are IRE sequences located at the 3′-UTR of TfR1 mRNA. Under iron deficiency condition, IRPs bind to the IRE sequence of TfR1 mRNA to prevent TfR1 mRNA degradation, which leads to the increase in TfR1 protein expression. On the other hand, under iron overload status, IRP is dissociated from IRE of TfR1's mRNA, bound to iron, which leads to the degradation of TfR1 mRNA by nuclease, and subsequently decreases TfR1 protein synthesis [23].

Divalent metal transporter 1 (DMT1) is a bivalent metal ion transporter with a molecular weight of approximately 61 kDa [24]. In 1997, both Harvard University and Yale University reported this transmembrane iron transporter [24, 25]. DMT1 plays a key role in maintaining the homeostasis of bivalent metal ions in cells, especially iron and manganese [26]. DMT1 can be divided into DMT1 (+IRE) and DMT1 (−IRE) according to whether its 3′-UTR contains IRE sequence. DMT1 (+IRE) is regulated by IRPs in response to intracellular iron level [18]. In the case of iron overload, IRPs bind to iron, which reduces the stability of DMT1 (+IRE) mRNA, promotes its degradation and decreases the level of DMT1 protein. On the contrary, under iron deficiency status, DMT1 expression increases [18].

2.3.3 Iron Storage Protein-Ferritin and FtMt

Ferritin, a major intracellular iron storage protein, consists of 24 subunits, a spherical polymer with a huge cavity, which can hold up to 4500 iron atoms [27]. Ferritin can be divided into two monomer subunits: heavy chain (H-ferritin, 21 kDa) and light chain (L-ferritin, 19 kDa), respectively [5, 27]. The H-ferritin subunit has the activity of ferrous oxidase, which oxidizes Fe^{2+} to Fe^{3+}. L-ferritin promotes the formation of iron nuclei and has a larger iron storage capacity than H-ferritin. Ferritin mainly participates in iron storage and utilization in cells, which limits the unstable labile iron pool (LIP) in cytoplasm and nucleus, protecting cells from iron-induced damage [5, 27].

The translation levels of both L-ferritin and H-ferritin are regulated by intracellular IRPs. There are IRE sequences located at the 5′-UTR of ferritin mRNAs. When the cell is at iron deficiency status, IRPs bind to IREs to prevent ferritin translation and downregulate ferritin expression, thereby reducing iron storage. Conversely, when an iron overload occurs in cells, IRPs cannot bind to IREs, the inhibition of ferritin translation is relieved, the expression of ferritin increases, and the capacity of iron storage in cells increases [28].

Mitochondrial ferritin (FtMt) is a kind of iron storage protein located in the mitochondrion, which has high homology with H-ferritin and can accumulate specifically in the mitochondrion [29]. It was found that FtMt could regulate iron distribution between cytoplasm and mitochondria by incorporating free iron in mitochondria, which reduces iron content in the cytoplasm and thereby reduces the production of ROS [30, 31]. FtMt is mainly expressed in mitochondria of some tissues with high oxygen consumption, such as testis, central nervous system (CNS), etc. [32]. This tissue-specific distribution may be related to the protection of mitochondria from iron-dependent oxidative damage caused by high metabolism and high oxygen consumption [33]. FtMt is also expressed in the brain, and the level of FtMt in neurons was found higher than that in glial cells [34].

2.3.4 *Iron Efflux Protein-FPN1*

Ferroportin (FPN1), also known as MPT1 or IREG1, is the currently only known multitransmembrane iron export protein in mammalian cells, with a molecular weight of about 62 kDa [6]. Three groups independently discovered this iron transporter in 2000 [35–37]. FPN1 commonly expresses on the membrane of all the cells, which allows Fe^{2+} to pass through. The basal surface of epithelial cells, the cellular surface of macrophages in liver and spleen, and trophoblast of placenta express relatively higher levels of FPN1, in order to export iron out of intestinal enterocytes, hepatocytes, and splenic macrophages to the blood. FPN1 transports ferrous ions out of cells with the assistance of ferrous oxidase ceruloplasmin (CP) or hephaestin (HP) [38]. The expression of FPN1 is mainly regulated by two aspects. On the one hand, IRPs regulate FPN1 expression at the post-transcriptional level via the IRP–IRE system. The 5′-UTR of FPN1 mRNA has the IRE structure [39]. IRP–IRE binding weakens the expression of FPN1 protein through translation inhibition. In addition, FPN1 protein level is also regulated by hepcidin. Hepcidin can recognize and bind to FPN1 on the membrane, which induces its internalization [14, 40]. After internalization, FPN1 and hepcidin are transported to lysosomes and both were degraded [41]. It was recently reported that hepcidin can also inhibit the activity of FPN1 [42, 43], which then reduces the release of intracellular iron, thus causing changes in intracellular or tissue iron levels. In addition, the regulation of hepcidin on FPN1 may also coordinate with IRPs, together affecting FPN1's level [44].

2.3.5 *Iron Metabolism Regulatory Proteins–IRPs and Hepcidin*

Iron regulatory proteins (IRPs) are cytosolic soluble RNA-binding proteins that bind to IRE sequences on mRNAs of iron transporters and iron storage proteins, regulating their expression [17, 45]. IRPs have two forms, iron regulatory protein 1 (IRP1, ~90 kDa) and iron regulatory protein 2 (IRP2, ~105 kDa). IRP1 is rich in most tissues, while IRP2 is most abundant in the brain and small intestine. When iron is deficient in cells, IRPs bind to IRE of the target mRNA, and while iron level is elevated, IRPs dissociate from IRE, regulating protein expression at the post-transcriptional level [17, 23]. IRPs bound to IREs of TfR1 mRNAs, as well as DMT1 mRNAs, promote their expression level, thus raising cellular iron absorption. Meanwhile, IRPs bound to IREs of FPN1 mRNAs reduce cellular iron release by suppressing FPN1's translation [39, 45]. Besides, IRPs regulate H-ferritin and L-ferritin translation, thus adjusting iron storage. These mechanisms have probably evolved to maintain the cytoplasmic LIP level at a steady level.

Hepcidin, a 25 amino acids peptide, produced mainly by the liver in response to high serum iron, is the 'master' regulator of iron homeostasis at systemic level [14]. Hepcidin was identified as a urinary antimicrobial peptide synthesized in the liver in 2000 [10]. Hepcidin restricts the release of iron by controlling FPN1 on the membrane, inducing its internalization and degradation [12]. Recently, the study reported that hepcidin can also downregulate the activity of FPN1 [42]. Hepcidin level is in turn regulated by body iron demand, through BMP/SMAD and TfR2 pathways. When the body iron is replete, a higher hepcidin concentration reduces iron absorption at intestine and impairs iron release from iron stores; but when the body iron is deficient, hepcidin level is low, thereby favoring iron absorption and delivery to the plasma from storage sites [15]. Inflammation is also an important factor affecting the expression of hepcidin. Interleukin-6 (IL-6) upregulates the expression of hepcidin through the STAT3 pathway [46]. In addition, hepcidin is negatively regulated by erythropoietin, oxygen environment, and sex hormones [16, 47].

2.4 Pathophysiology of Iron

2.4.1 Iron Deficiency

Iron deficiency is a state that the body iron amount is lower than the normal physiologic conditions, and the body lacks enough iron to supply its needs. Iron deficiency may impair the synthesis of essential iron-containing proteins that are required for normal cellular physiology and result in a range of adverse consequences [48]. When the loss of iron is not sufficiently compensated by adequate iron intake from the diet, a state of iron deficiency will develop over time [48].

Untreated iron deficiency can lead to iron deficiency anemia, the most common type of anemia [49]. In iron deficiency anemia, the body lacks sufficient amounts of iron to produce the protein hemoglobin, thereby resulting in inadequate red blood cells (RBCs). Besides, iron deficiency can also be resulted from genetic disturbances in iron homeostasis. In particular, mutations in the TMPRSS6 gene, which encodes an upstream regulator of hepcidin, can lead to iron-refractory iron deficiency anemia [50]. The treatment of mild iron deficiency can be achieved simply by increasing the amount of bioavailable iron in the diet or by taking iron supplements; while in the severe case of iron deficiency anemia, an intravenous iron infusion may be used to efficiently deliver a large amount of iron [51].

The anemia of chronic disease is the second most common form of anemia that is caused by iron deficiency with inflammation [52]. The anemia of chronic disease is usually due to the presence of chronic infection, chronic immune activation, malignancy, or various chronic inflammatory states. These conditions all produce massive elevation of pro-inflammatory cytokine IL-6. IL-6 stimulates hepcidin production and release from the liver, which in turn reduces the iron efflux protein FPN1, resulting in reduced iron uptake at intestine and reduced iron release from iron stores and macrophages, thereby reducing iron in the circulation. The plasma iron concentration decreases, and macrophages and other cell types sequester iron. If this condition persists, the iron supply to the erythroid marrow can be compromised, resulting in reduced erythropoiesis, namely the anemia of chronic disease [52]. Hepcidin antagonists or agents that block hepcidin gene synthesis and subsequently increase intestinal iron absorption have shown promise in the treatment of the anemia of inflammation [53].

2.4.2 Iron Overload

Iron overload is also associated with a variety of diseases [54]. Excess LIP in the cell may catalyze reactions that generate ROS and consequently cause oxidative damage to cells and tissues, which can finally lead to tissue fibrosis and organ dysfunctions [55]. Some iron overload diseases are caused by genetic mutations in the genome, such as hemochromatosis, atransferrinemia, aceruloplasminemia, and Friedreich ataxia [54–56]. Some are resulted from the chronic disorders, e.g., chronic liver disease [57]. Iron overload is also complicated with a range of other diseases, such as cystic fibrosis, fatty liver disease, and neurodegenerative diseases, including Alzheimer's disease (AD), Parkinson's disease (PD), and so on [54, 58].

Hemochromatosis is a major group of primary iron overload diseases, which is resulted from mutations in genes that are involved in the transport of iron or the regulation of iron metabolism [59]. Currently, there are five main types of hemochromatosis are identified due to mutations in the genes encoding hemochromatosis protein *HFE*, hemojuvelin (HJV), hepcidin, TfR2, and FPN1 [59, 60]. The common feature of these different types is a reduced hepcidin expression level [59]. The reduced hepcidin level leads to a higher FPN1 level on the membrane of intestinal enterocytes, which increases the absorption of dietary iron, and therefore the body iron load increases. In these patients, iron accumulation was observed in many organs, particularly in the liver, heart, and pancreas [54]. Clinical syndromes of the hemochromatosis patients include hepatic fibrosis and cirrhosis, hepatocellular carcinoma, arthritis, cardiomyopathy, and so on [59]. The disease of hemochromatosis is relatively common in the populations of Northern European origin, but is a rare case in other populations [60]. Hemochromatosis is typically treated via phlebotomy with the removal of ~0.5 g Fe/L blood, which is the currently most effective and relatively inexpensive method [54]. After removal, the serum iron and RBCs in the blood, iron in the storage sites, notably in the liver, will be subsequently released to replenish the reduced RBCs in blood for erythropoiesis [54].

Thalassemia is one kind of anemia with iron overload instead of iron deficiency. In thalassemia, patients have defects in either the α- or β-globin chain, causing the production of abnormal hemoglobin, thereby characterized by reduced RBCs and thus anemia [61]. However, blood transfusions for treatment of this inherited disorder usually lead to secondary iron overload, with resulting heart or liver diseases, infections, and osteoporosis. It is most common among people of Italian, Greek, Middle Eastern, South Asian, and African descent. In the treatment of iron overload in thalassemias, iron chelators are commonly used [61]. Hepcidin mimetics or agents that increase hepcidin expression could also be promising drugs in the treatment of hemochromatosis and thalassemia [62].

Abnormal iron accumulation has been detected in various neurological diseases in CNS, including neurodegeneration with brain iron accumulation (NBIA) diseases, AD, PD, Friedreich's ataxia, amyotrophic lateral sclerosis, stroke, and so on [63, 64]. Although the causes of most of the neurological disorders have not been fully understood, common features of these diseases are iron accumulation and increased oxidative damage [63]. Iron chelators, such as deferoxamine (DFO), have been shown good efficacy as potential therapeutic agents for iron overload-associated neurological disorders [65]. As early as in 1991, McLachlan et al. had used DFO to treat AD in a clinical trial. They found that continuous administration of DFO could alleviate AD-induced cognitive impairment [66]. In the recent years, it was found that DFO treatment reduced the cognitive decline and the accumulation of amyloid-β in APP/PS1 mice, which may be related to the induction of M2 activation and inhibition of M1 activation by DFO in the hippocampus of mice [67]. Researchers have successfully developed a brain-targeting peptide-modified nanopolymer encapsulated DFO, which can cross the blood–brain barrier through receptor-mediated endocytosis [68]. This improvement greatly increased the efficiency of DFO entering the brain, and significantly decreased the iron content in substantia nigra and striatum, reduced oxidative stress level, and alleviated dopaminergic neuron damage in PD mouse model [68].

2.5 Conclusions

Iron is an essential trace element in human body, but it is also toxic in excess. Thus, iron homeostasis at cellular level and the whole-body iron concentration are tightly regulated to keep it within the optimal physiologic range. Body iron is acquired from the diet and exported into the circulation for utilization. Therefore, the amount of iron uptake at intestinal enterocyte is fairly critical. Both the levels of DMT1 and FPN1 expressed by intestinal enterocytes are must under precise regulations. The liver-derived peptide hepcidin in serum plays a major regulatory in controlling FPN1 level on the membrane of intestinal enterocytes. While intracellularly, IRPs regulate the levels of DMT1, TfR1, ferritin and FPN1 in response to iron level changes. Therefore, an adequate and steady iron supplement is warranted for the utilization of cells around the body. Similar as the iron regulation in enterocyte, the cellular iron levels in other cells and tissues are controlled differently by different cell types due to the expression of particular iron regulatory and transport proteins. Abnormal levels of body iron, too little or too much, are commonly associated with diseases of iron deficiency or iron overload. Despite major recent advances have achieved regarding the mechanisms of cellular iron metabolism, much remains to be learned about iron physiology and pathophysiology.

References

1. Muckenthaler MU, Rivella S, Hentze MW, Galy B (2017) A red carpet for iron metabolism. Cell 168(3):344–361
2. Anderson GJ, Wang F (2012) Essential but toxic: controlling the flux of iron in the body. Clin Exp Pharmacol Physiol 39(8):719–724
3. Aisen P, Leibman A, Zweier J (1978) Stoichiometric and site characteristics of the binding of iron to human transferrin. J Biol Chem 253(6):1930–1937
4. Huebers HA, Finch CA (1987) The physiology of transferrin and transferrin receptors. Physiol Rev 67(2):520–582
5. Harrison PM, Arosio P (1996) The ferritins: molecular properties, iron storage function and cellular regulation. Biochim Biophys Acta 1275(3):161–203
6. Drakesmith H, Nemeth E, Ganz T (2015) Ironing out ferroportin. Cell Metab 22(5):777–787
7. Paoli M, Marles-Wright J, Smith A (2002) Structure-function relationships in heme-proteins. DNA Cell Biol 21(4):271–280
8. Hentze MW, Muckenthaler MU, Galy B, Camaschella C (2010) Two to tango: regulation of Mammalian iron metabolism. Cell 142(1):24–38
9. Aisen P (1992) Entry of iron into cells: a new role for the transferrin receptor in modulating iron release from transferrin. Ann Neurol 32(Suppl):S62–S68
10. Park CH, Valore EV, Waring AJ, Ganz T (2001) Hepcidin, a urinary antimicrobial peptide synthesized in the liver. J Biol Chem 276(11):7806–7810
11. Nicolas G, Chauvet C, Viatte L, Danan JL, Bigard X, Devaux I et al (2002) The gene encoding the iron regulatory peptide hepcidin is regulated by anemia, hypoxia, and inflammation. J Clin Invest 110(7):1037–1044
12. Ramey G, Deschemin JC, Durel B, Canonne-Hergaux F, Nicolas G, Vaulont S (2010) Hepcidin targets ferroportin for degradation in hepatocytes. Haematologica 95(3):501–504
13. Beutler E (2004) Cell biology. "Pumping" iron: the proteins. Science 306(5704):2051–2053
14. Nemeth E, Tuttle MS, Powelson J, Vaughn MB, Donovan A, Ward DM et al (2004) Hepcidin regulates cellular iron efflux by binding to ferroportin and inducing its internalization. Science 306(5704):2090–2093
15. Parrow NL, Fleming RE (2014) Bone morphogenetic proteins as regulators of iron metabolism. Annu Rev Nutr 34(34):77–94
16. Ganz T (2011) Hepcidin and iron regulation, 10 years later. Blood 117(17):4425–4433
17. Rouault TA (2006) The role of iron regulatory proteins in mammalian iron homeostasis and disease. Nat Chem Biol 2(8):406–414
18. Leipuviene R, Theil EC (2007) The family of iron responsive RNA structures regulated by changes in cellular iron and oxygen. Cell Mol Life Sci 64(22):2945–2955
19. Meyron-Holtz EG, Ghosh MC, Rouault TA (2004) Mammalian tissue oxygen levels modulate iron-regulatory protein activities in vivo. Science 306(5704):2087–2090
20. Anderson GJ, Frazer DM (2017) Current understanding of iron homeostasis. Am J Clin Nutr 106(Suppl 6):1559S–1566S
21. Bali PK, Zak O, Aisen P (1991) A new role for the transferrin receptor in the release of iron from transferrin. Biochemistry 30(2):324–328
22. Cheng Y, Zak O, Aisen P, Harrison SC, Walz T (2004) Structure of the human transferrin receptor-transferrin complex. Cell 116(4):565–576
23. Wallander ML, Leibold EA, Eisenstein RS (2006) Molecular control of vertebrate iron homeostasis by iron regulatory proteins. Biochim Biophys Acta 1763(7):668–689
24. Gunshin H, Mackenzie B, Berger UV, Gunshin Y, Romero MF, Boron WF et al (1997) Cloning and characterization of a mammalian proton-coupled metal-ion transporter. Nature 388(6641):482–488
25. Fleming MD, Trenor CC, Su MA, Foernzler D, Beier DR, Dietrich WF et al (1997) Microcytic anaemia mice have a mutation in Nramp2, a candidate iron transporter gene. Nat Genet 16(4):383–386

26. Garrick MD, Dolan KG, Horbinski C, Ghio AJ, Higgins D, Porubcin M et al (2003) DMT1: a mammalian transporter for multiple metals. Biometals 16(1):41–54
27. Theil EC (1987) Ferritin: structure, gene regulation, and cellular function in animals, plants, and microorganisms. Annu Rev Biochem 56:289–315
28. Arosio P, Ingrassia R, Cavadini P (2009) Ferritins: a family of molecules for iron storage, antioxidation and more. Biochim Biophys Acta 1790(7):589–599
29. Levi S, Corsi B, Bosisio M, Invernizzi R, Volz A, Sanford D et al (2001) A human mitochondrial ferritin encoded by an intronless gene. J Biol Chem 276(27):24437–24440
30. Corsi B, Cozzi A, Arosio P, Drysdale J, Santambrogio P, Campanella A et al (2002) Human mitochondrial ferritin expressed in HeLa cells incorporates iron and affects cellular iron metabolism. J Biol Chem 277(25):22430–22437
31. Nie G, Sheftel AD, Kim SF, Ponka P (2005) Overexpression of mitochondrial ferritin causes cytosolic iron depletion and changes cellular iron homeostasis. Blood 105(5):2161–2167
32. Santambrogio P, Biasiotto G, Sanvito F, Olivieri S, Arosio P, Levi S (2007) Mitochondrial ferritin expression in adult mouse tissues. J Histochem Cytochem 55(11):1129–1137
33. Cocco E, Porrini V, Derosas M, Nardi V, Biasiotto G, Maccarinelli F et al (2013) Protective effect of mitochondrial ferritin on cytosolic iron dysregulation induced by doxorubicin in HeLa cells. Mol Biol Rep 40(12):6757–6764
34. Wang L, Yang H, Zhao S, Sato H, Konishi Y, Beach TG et al (2011) Expression and localization of mitochondrial ferritin mRNA in Alzheimer's disease cerebral cortex. PLoS ONE 6(7):e22325
35. Abboud S, Haile DJ (2000) A novel mammalian iron-regulated protein involved in intracellular iron metabolism. J Biol Chem 275(26):19906–19912
36. Donovan A, Brownlie A, Zhou Y, Shepard J, Pratt SJ, Moynihan J et al (2000) Positional cloning of zebrafish ferroportin1 identifies a conserved vertebrate iron exporter. Nature 403(6771):776–781
37. McKie AT, Marciani P, Rolfs A, Brennan K, Wehr K, Barrow D et al (2000) A novel duodenal iron-regulated transporter, IREG1, implicated in the basolateral transfer of iron to the circulation. Mol Cell 5(2):299–309
38. Knutson MD, Oukka M, Koss LM, Aydemir F, Wessling-Resnick M (2005) Iron release from macrophages after erythrophagocytosis is up-regulated by ferroportin 1 overexpression and down-regulated by hepcidin. Proc Natl Acad Sci USA 102(5):1324–1328
39. Lymboussaki A, Pignatti E, Montosi G, Garuti C, Haile DJ, Pietrangelo A (2003) The role of the iron responsive element in the control of ferroportin1/IREG1/MTP1 gene expression. J Hepatol 39(5):710–715
40. Qiao B, Sugianto P, Fung E, del-Castillo-Rueda A, Moran-Jimenez MJ, Ganz T et al (2012) Hepcidin-induced endocytosis of ferroportin is dependent on ferroportin ubiquitination. Cell Metab 15(6):918–924
41. Preza GC, Pinon R, Ganz T, Nemeth E (2013) Cellular catabolism of the iron-regulatory peptide hormone hepcidin. PLoS ONE 8(3):e58934
42. Zhang DL, Rouault TA (2018) How does hepcidin hinder ferroportin activity? Blood 131(8):840–842
43. Zhang DL, Wu J, Shah BN, Greutelaers KC, Ghosh MC, Ollivierre H et al (2018) Erythrocytic ferroportin reduces intracellular iron accumulation, hemolysis, and malaria risk. Science 359(6383):1520–1523
44. Wang L, Liu X, You LH, Ci YZ, Chang S, Yu P et al (2019) Hepcidin and iron regulatory proteins coordinately regulate ferroportin 1 expression in the brain of mice. J Cell Physiol 234(5):7600–7607
45. Pantopoulos K (2004) Iron metabolism and the IRE/IRP regulatory system: an update. Ann N Y Acad Sci 1012:1–13
46. Nemeth E, Valore EV, Territo M, Schiller G, Lichtenstein A, Ganz T (2003) Hepcidin, a putative mediator of anemia of inflammation, is a type II acute-phase protein. Blood 101(7):2461–2463
47. Vokurka M, Krijt J, Sulc K, Necas E (2006) Hepcidin mRNA levels in mouse liver respond to inhibition of erythropoiesis. Physiol Res 55(6):667–674
48. Zimmermann MB, Hurrell RF (2007) Nutritional iron deficiency. Lancet 370(9586):511–520

49. Naigamwalla DZ, Webb JA, Giger U (2012) Iron deficiency anemia. Can Vet J-Revue Vet Can 53(3):250–256
50. Finberg KE, Heeney MM, Campagna DR, Aydinok Y, Pearson HA, Hartman KR et al (2008) Mutations in TMPRSS6 cause iron-refractory iron deficiency anemia (IRIDA). Nat Genet 40(5):569–571
51. Powers JM, Buchanan GR (2014) Diagnosis and management of iron deficiency anemia. Hematol Oncol Clin North Am 28(4):729–745, vi–vii
52. Peng YY, Uprichard J (2017) Ferritin and iron studies in anaemia and chronic disease. Ann Clin Biochem 54(1):43–48
53. Camaschella C (2017) New insights into iron deficiency and iron deficiency anemia. Blood Rev 31(4):225–233
54. Fleming RE, Ponka P (2012) Iron overload in human disease. N Engl J Med 366(4):348–359
55. Anderson GJ (2007) Mechanisms of iron loading and toxicity. Am J Hematol 82(12):1128–1131
56. Doyle A, Rusli F, Bhathal P (2015) Aceruloplasminaemia: a rare but important cause of iron overload. BMJ Case Rep 2015
57. Kew MC (2014) Hepatic iron overload and hepatocellular carcinoma. Liver Cancer 3(1):31–40
58. Ward RJ, Zucca FA, Duyn JH, Crichton RR, Zecca L (2014) The role of iron in brain ageing and neurodegenerative disorders. Lancet Neurol 13(10):1045–1060
59. Pietrangelo A (2010) Hereditary hemochromatosis: pathogenesis, diagnosis, and treatment. Gastroenterology 139(2):393–408, 408 e391–e392
60. Barton JC, Edwards CQ, Acton RT (2015) HFE gene: structure, function, mutations, and associated iron abnormalities. Gene 574(2):179–192
61. Taher AT, Weatherall DJ, Cappellini MD (2018) Thalassaemia. Lancet 391(10116):155–167
62. Olivieri NF, Brittenham GM (2013) Management of the thalassemias. Cold Spring Harb Perspect Med 3(6)
63. Ndayisaba A, Kaindlstorfer C, Wenning GK (2019) Iron in neurodegeneration—cause or consequence? Front Neurosci 13:180
64. Rouault TA (2013) Iron metabolism in the CNS: implications for neurodegenerative diseases. Nat Rev Neurosci 14(8):551–564
65. Dusek P, Schneider SA, Aaseth J (2016) Iron chelation in the treatment of neurodegenerative diseases. J Trace Elem Med Biol 38:81–92
66. Crapper McLachlan DR, Dalton AJ, Kruck TP, Bell MY, Smith WL, Kalow W et al (1991) Intramuscular desferrioxamine in patients with Alzheimer's disease. Lancet 337(8753):1304–1308
67. Zhang Y, He ML (2017) Deferoxamine enhances alternative activation of microglia and inhibits amyloid beta deposits in APP/PS1 mice. Brain Res 1677:86–92
68. You L, Wang J, Liu T, Zhang Y, Han X, Wang T et al (2018) Targeted brain delivery of rabies virus glycoprotein 29-modified deferoxamine-loaded nanoparticles reverses functional deficits in parkinsonian mice. ACS Nano 12(5):4123–4139

Chapter 3
Brain Iron Metabolism and Regulation

Peng Yu and Yan-Zhong Chang

Abstract With the development of research, more and more evidences suggested that mutations in the genes associated with brain iron metabolism induced diseases in the brain. Brain iron metabolism disorders might be one cause of neurodegenerative diseases. This review mainly summarizes the normal process of iron entry into the brain across the blood–brain barrier, and the distribution and transportation of iron among neurons and glial cells, as well as the underlying regulation mechanisms. To understand the mechanisms of iron metabolism in the brain will provide theoretical basis to prevent and cure brain diseases related to iron metabolism disorders.

Keywords Blood–brain barrier · Transferrin receptor · Ferroportin 1 · Iron regulatory protein · Hepcidin

3.1 Introduction

Iron is one of the most abundant trace elements and plays great roles in keeping physiological homeostasis in the body. In the central nervous system (CNS), iron participates in the synthesis of myelin and neurotransmitters [24, 39]. Iron deficiency in the brain of infants will lead to abnormal neural development and mental retardation [21, 70]. While iron accumulation in specific brain regions is also shown to be related to the patients with neurodegenerative diseases such as Parkinson's Disease (PD), Alzheimer's Disease (AD) and Huntington's Disease (HD). Excess iron can react with cellular H_2O_2 and generate reactive oxygen species (ROS) through Fenton

---

P. Yu (✉) · Y.-Z. Chang (✉)
Laboratory of Molecular Iron Metabolism, Key Laboratory of Animal Physiology, Biochemistry and Molecular Biology of Hebei Province, College of Life Sciences, Hebei Normal University, 20, Nanerhuan Eastern Road, Shijiazhuang, Hebei Province 050024, China
e-mail: yupeng0311@163.com

Y.-Z. Chang
e-mail: chang7676@163.com

Y.-Z. Chang (ed.), *Brain Iron Metabolism and CNS Diseases*,
Advances in Experimental Medicine and Biology 1173,
https://doi.org/10.1007/978-981-13-9589-5_3

reaction, and increased ROS could induce a cascade of events to destroy membrane, protein, and nucleic acid, which caused apoptosis and even cell death [38, 65, 96]. Therefore, it is very important to keep iron homeostasis in the brain.

3.2 Iron Uptake into the Brain

3.2.1 Iron Enters the Brain Across Blood–Brain Barrier (BBB)

Iron homeostasis in the brain was regulated strictly in mammalian animals, and it depends on BBB and many molecules involved in the transportation of iron from the blood circulation. In CNS, the microvascular endothelial cell (BMVEC), together with astrocyte end-feet and pericytes, forms BBB which is the primary gatekeeper for iron entry into the brain. Because there are tight junctions between BMVEC, iron was first uptake into BMVEC from blood and then across BBB with the help of many iron metabolism-related proteins [1, 44].

Long time ago, it is thought that iron can cross BBB into the brain mainly during infancy before the formation of BBB. Now, new data demonstrated that iron can also enter the brain across BBB in the adult brain, which depends on transferrin receptor 1 (TfR1) [56] and ferroportin 1 (FPN1) [50, 71, 92]. Iron can be bound in transferrin (Tf) and transport in the blood, and Tf-TfR1 has been known as the main major route of iron transport across the luminal membrane of the capillary endothelium from the blood into endothelial cells [8, 47, 56]. First, Tf-bound iron in serum reached BBB, the Tf-Fe binds to TfR1 in BMVEC, which forms endosomes after endocytosis of Tf-TfR1. When pH value decreased to 5.5~6.5 in the endosomes, iron dissociated from Tf and reduced to ferrous iron which released from the endosomes into endothelial cytoplasm with the help of divalent metal transporter 1 (DMT1) [80, 81]. After dissociation of iron from Tf-TfR1, apo-Tf-TfR1 returned to the apical surface of BMVEC for the next round of iron uptake [9, 74], and intracellular iron is then either incorporated into functional proteins, stored in ferritin, or continued to transport across the abluminal membrane of the endothelial cell into the brain interstitium and parenchyma [8, 56]. During the process of iron transport across the basal membrane of intestinal epithelial cells, ferroportin 1 (FPN1) is an iron exporter and responsible for iron efflux from epithelia into the blood in the presence of hephaestin (HP) and ceruloplasmin (CP). Our previous results showed that FPN1 was distributed in murine brain such as cerebral cortex, hippocampus, and striatum [88]. Others also reported that FPN1 was also expressed in BBB, ependymocytes and choroid plexus, and that FPN1 was especially expressed in the basal surface of BMVEC [90]. Moreover, CP was also mainly expressed in the end-feet of astrocytes surrounding microvascular vessels in the brain [41, 42, 68]. These evidences suggested that FPN1/CP might play critical roles in iron efflux into brain parenchyma from BMVEC just similar to that of enterocytes in duodenum. It is also reported that astrocytes can also directly uptake ferrous iron from BMVEC through the end-feet [49, 64]. However, there are no direct evidences in vivo.

Besides Tf-TfR1 dependent iron uptake pathway, there are Tf-independent iron transport ways. Ferritin (H-ferritin) can also deliver iron to the brain across BBB, which is an iron transport system independent of Tf-TfR system. Ferritin might bind to ferritin receptor or be transcytosed across the BBB [24]. Lactoferrin (Lf) and its receptor (LfR), glycosylphosphatidylinositol anchored form of melanotransferrin (p97, MTf), and soluble form of MTf also play roles in iron uptake across BBB from the blood [22, 59]. Lf belongs to the Tf family and it is an iron-binding glycoprotein which is distributed in milk or other secretion in peripheral tissues. In CNS, Lf is expressed and secreted from activated microglia [2, 23], and LfR is expressed in neurons, BMVEC, and some glial cells. Lf-LfR might play similar roles as Tf-TfR1 in iron transport into the brain parenchyma [21]. MTf is a transferrin homolog, and GPI-MTf also plays similar roles as Tf-TfR1 in iron transport across BBB, and secreted MTf seems to play more important roles in iron metabolism of patients [59].

However, it is still not fully understood the detailed regulatory mechanisms of the crosstalk between brain iron uptake and release at the BBB in vivo.

3.2.2 Iron Enters the Brain Across Blood Cerebrospinal Fluid Barrier (BCB)

It is speculated that iron cross the blood–cerebrospinal fluid barrier (BCB) from the blood is similar to iron transport at BBB. In situ hybridization found that the choroid plexus expressed high levels of TfR1, DMT1, FPN1, Dcytb, and soluble form of CP and HP, which suggested that choroid plexus might mediate significant iron transport in CNS [52, 67, 78]. The choroid plexus can synthesize and secret Tf, which might be involved in iron transport from BCB to the brain. But choroid plexus-derived Tf does not seem too important to bind iron and transport iron in the brain interstitium because [^{59}Fe^{125}I]Tf injected intracerebroventricularly was never observed in regions distant from the CSF. Tf in the CSF probably plays a significant role in export to the blood of iron [55].

Stromal cell-derived receptor 2 (SDR2) is expressed in the choroid plexus and ependymal cells lining the four ventricles with in situ hybridization analysis [85]. SDR2 is the homolog of duodenal cytochrome b that is a ferric reductase, and DMT1 is also located in the choroid plexus [29], so SDR2 might also have ferric reductase activity and participate in iron uptake with DMT1 at BCB.

3.3 The Distribution and Transportation of Iron in the Brain

3.3.1 Iron Is Distributed Unevenly in Different Brain Regions

After iron is released into brain interstitial fluid, it is distributed and transported among different cell types in different brain regions. It is reported that 10% of non-heme iron entered the brain across BBB from the vertebrate blood and that iron levels are different in different brain regions [11, 77]. Iron contents are most abundant in the extrapyramidal system, especially in the basal ganglia region [58, 60], and iron concentrations are high in subtantia nigra (SN), red nucleus, and cerebellar dentate gyrus, while iron is relatively low in cerebral cortex, and iron levels are the lowest in white matter and medulla oblongata [17, 58, 60].

3.3.2 Distribution and Transportation of Iron in Different Cell Types of the Brain

Iron is released into the brain interstitial fluid across BBB, and it is absorbed by neurons and different types of glial cells, but iron is distributed unevenly in these cells [57, 77]. Neuroglia expressed almost all the iron transporters and distinct profiles of iron metabolism related proteins [6, 32, 78, 89]. Data from primary cultures of neurons, astrocytes, and microglia showed that iron retention in microglia was three times higher than that of neurons when they were incubated with iron [7]. Studies in a hippocampal slice culture mode also supported that ferritin mRNA was induced by iron predominantly in microglia and oligodendrocytes, followed by astrocytes, but never in neurons [31]. So neuroglia are indispensable to maintain iron homeostasis and normal physiological function in the brain.

As for neurons, they express TfR1 and acquire iron through classical Tf-TfR1-dependent iron uptake pathway from the brain interstitial fluid [30]. Neurons can also uptake ferrous and ferric iron through DMT1 and trivalent cation-specific transporter (TCT1), respectively [4, 39, 73].

As for astrocytes, the major glial cell type in CNS, they are important components of BBB, so they play crucial roles in iron transport and are responsible for iron distribution in the brain [10, 57]. They exclusively express glycosylphosphatidylinositol-anchored CP (GPI-CP) in the end-feet surrounding BBB [35, 67]. Astrocytes also secrete a soluble form of CP into interstitium which also facilitates iron trafficking across BBB into the brain [52, 83]. Both GPI-CP and soluble form of CP are required for the stability of FPN1 in the membrane and iron release from BMVEC. Moreover, hephaestin present in astrocytes [10, 86] can also oxidize BMVEC released ferrous iron to ferric iron so as to facilitate incorporation of iron into Tf and the following transportation [13, 79, 94]. Astrocytes acquire iron across the interstitial space using

NTBI mechanisms such as citrate, ATP, ascorbic acid, DMT1, and Zip [45, 69, 91]. They release iron through FPN1, and the released iron can be acquired by neurons and other types of glial cells.

Oligodendrocytes are unique to the brain, because they are the only cells with synthesized Tf and releasing Tf, and they express the Tim2 (T cell immunoglobulin and mucin domain protein-2) ferritin transporter and contain a large amount of the brain's lipids and cholesterol [70], and they have high levels of iron and require large amounts of iron in the myelination of neuronal axons to form the white matter in the brain [5].

Microglia compromise approximately 12% of cells in the brain, and the resident microglia are very stable. Microglia express the transporters or molecules involved in iron metabolism, and they are most efficient in accumulating iron as mentioned above [7, 83]. Under inflammation and environmental and endogenous stimuli, microglia can sense and recognize the changes, and then, they become activated, iron retention and affect the iron metabolism and function of neurons [83, 84]. Immunofluorescence and hybridization in situ results revealed that activated microglia could synthesize and secrete lactoferrin [2, 23], which can also affect the iron metabolism of lactoferrin receptor expressing cells through the receptor-mediated pathway [91].

After iron is used for neuronal and glial metabolism and stored in ferritin, excess iron is released via FPN1. The excess iron can be exported back to the interstitial fluid or released into the cerebrospinal fluid in the brain ventricles [8, 27, 97], and the choroidal epithelia capture iron by TfR1-dependent or TfR1-independent pathway and then transport it back to the blood circulation [27].

3.4 The Regulation of Iron Metabolism in the Brain

Iron homeostasis is very important for normal brain functions. Iron homeostasis depends on iron uptake, storage, and release [34]. Iron regulatory proteins (IRPs) and hepcidin are critical regulators and coordinate with each other to maintain iron homeostasis [28, 87].

3.4.1 IRPs Regulate Brain Iron Metabolism from Cellular Level

IRP1 share high sequence homology with IRP2, and IRPs will regulate iron uptake, iron storage, and iron release proteins with iron responsive element (IRE) in the gene transcripts [61]. When cellular iron levels change, IRPs will regulate iron metabolism through IRE-IRP regulatory systems at post-transcription level [40, 66].

In 2006, Rouault lab reported that IRP1$^{-/-}$IRP2$^{-/-}$ double knockout mice were embryonically lethal [82], which demonstrated the critical roles of IRPs in life.

However, IRP1 and IRP2 have different distribution patterns, only IRP2 is highly expressed in the brain [54]. Rouault lab and others also generated IRP2$^{-/-}$ mice, and it was reported that excessive iron is accumulated in the brain of IRP2$^{-/-}$ mice and neurodegenerative diseases like impairments were observed in aging IRP2$^{-/-}$ mice [12, 25, 46, 66, 98], which further confirmed the crucial role of IRP2 in the regulation of brain iron metabolism.

When cellular iron concentrations change, the expression of IRP2 will be altered accordingly. Decreased cellular iron contents can increase the expression of IRP2, which can bind to IRE in 5′-untranslated region (UTR) of the mRNA such as ferritin mRNA, IRE in 3′UTR of transferrin receptor 1 (TfR1) mRNA, and DMT1 with IRE [DMT1 (+IRE)] mRNA, thus induce the inhibition of ferritin expression, elevation of TfR1 and DMT1 (+IRE) expression, which increase iron uptake and decrease iron storage to relieve the iron deficiency in the cell. The vice is versa [20, 54, 61, 66, 93]. Therefore, IRE-IRP system can maintain iron homeostasis responding to altered cellular iron levels in the brain.

3.4.2 *Hepcidin Regulates Brain Iron Metabolism from Systemic Level*

Hepcidin, an iron-regulatory hormone, plays an essential role in maintaining body iron homeostasis [33, 62, 63]. Hepcidin regulates organismal iron concentration and tissue iron distribution by controlling intestinal iron absorption, iron recycling by macrophages, and iron mobilization from hepatic stores [33, 43, 62]. In peripheral tissues, hepcidin is upregulated responding to inflammation [48] and iron overload in the blood; then, hepcidin binds to the extracellular loop of FPN1 and causes its internalization and degradation, thereby reducing cellular iron efflux from intestine enterocytes and macrophages, which relieves the iron overload status and decreases the iron levels to the normal range in the blood [3, 14, 16, 63, 75].

Recent studies have also revealed that hepcidin mRNA levels are different in different brain regions, and that hepcidin immunoreactivity can be detected in both neurons and GFAP-positive glial cells [95], which suggests that hepcidin can be also involved in the regulation of brain iron metabolism [19, 72, 88, 95]. FPN1-dependent iron export systems might also play a key role in iron transport across BBB similar to that in the abluminal membrane of enterocytes in the gut [15, 53].

Several studies have provided evidences of an important regulatory role for hepcidin in brain iron metabolism. It is reported that hepcidin mRNA levels increase with aging and injection of hepcidin peptide into the lateral cerebral ventricle decreases FPN1 levels, resulting in brain iron overload [88]. So hepcidin can respond to changes of systemic iron levels and inflammation and then regulates FPN1 to keep iron homeostasis [26]. In brain microvascular endothelial cells, it is demonstrated that hepcidin downregulates iron efflux from hBMVECs by internalization of FPN1 in vitro [51]. In primary cultures of astrocytes, extracellular hepcidin triggers the decreased expression of TfR1, DMT1 and FPN1, and hepcidin inhibits TfR1 expression via a cyclic AMP-protein kinase A pathway, which implies the existence of a novel hepcidin-receptor on the membrane of astroglia [18, 83]. However, there are no direct evidences to show that brain hepcidin regulates the content of brain iron through effects on FPN1 of BMVEC in vivo.

In fact, hepcidin mRNA levels in the brain have been shown to be much lower than those in non-neural tissues or blood by qPCR experiments [88], but most researches draw the conclusions of hepcidin-mediated regulation of FPN1 expression in BMVECs or neurons based on the usage of a considerably high concentration of supraphysiological hepcidin (approximately 20-fold higher than that in blood). Based on these data and facts, it is still to be elucidated how the low concentration of hepcidin in the brain can regulate FPN1 in BMVECs and other types of cells.

It is worth noting that hepcidin and IRE-IRP systems must have communications and coordination in their regulation of brain iron metabolisms [87]. In addition, erythroferrone (ERFE) is a newly identified protein by Ganz group, and ERFE could inhibit hepcidin and thus regulate iron metabolism in response to erythropoietin stimulation under the stress condition [36, 37]. ERFE is also detected in the brain with real-time PCR [37], but it is still unclear about its regulation of brain iron metabolism. It is also unknown if there is any other functions of ERFE in the brain, which needs to be further investigated.

3.5 Conclusion

Although brain iron metabolism and its regulation are complicated and there are still some puzzles to be unraveled, the regulation profile is becoming clearer for us. Iron is transported in the blood and entered the brain across BBB. Then iron is acquired by neurons and different types of glial cells through iron uptake proteins, and released from these cells through FPN1. Brain iron homeostasis is coordinately regulated by IRP2 and hepcidin at post-transcriptional and protein levels, respectively. However, there are still some problems to be answered and understood, such as how the iron was released from the brain? Is the recently concerned glymphatic pathway involved in it [76]? Which kind of signal initiates the uptake of iron and release from the brain? The answers to these problems will not only enrich the theories of brain iron metabolism and regulation, but also provide strategies to prevent and treat iron metabolism disorders and related neurodegenerative diseases.

Acknowledgements The work was supported by the National Natural Science Foundation of China (31520103908, 31970905), Hebei Provincial Natural Science Foundation (C2017205140), Hebei Provincial Education Department Foundation of China (ZD2015105), Hebei Provincial Fund for studying abroad and returning to China (C201862), and Hebei normal University Outstanding Youth Fund (L2018J05).

References

1. Abbott NJ, Ronnback L, Hansson E (2006) Astrocyte-endothelial interactions at the blood-brain barrier. Nat Rev Neurosci 7(1):41–53
2. An L, Sato H, Konishi Y, Walker DG, Beach TG, Rogers J, Tooyama I (2009) Expression and localization of lactotransferrin messenger RNA in the cortex of Alzheimer's disease. Neurosci Lett 452(3):277–280
3. Anderson GJ, Wang F (2012) Essential but toxic: controlling the flux of iron in the body. Clin Exp Pharmacol Physiol 39(8):719–724
4. Attieh ZK, Mukhopadhyay CK, Seshadri V, Tripoulas NA, Fox PL (1999) Ceruloplasmin ferroxidase activity stimulates cellular iron uptake by a trivalent cation-specific transport mechanism. J Biol Chem 274(2):1116–1123
5. Bartzokis G (2002) Schizophrenia: breakdown in the well-regulated lifelong process of brain development and maturation. Neuropsychopharmacology: Official Publication of the American College of Neuropsychopharmacology 27(4):672–683
6. Benarroch EE (2009) Brain iron homeostasis and neurodegenerative disease. Neurology 72(16):1436–1440
7. Bishop GM, Dang TN, Dringen R, Robinson SR (2011) Accumulation of non-transferrin-bound iron by neurons, astrocytes, and microglia. Neurotox Res 19(3):443–451
8. Bradbury MW (1997) Transport of iron in the blood-brain-cerebrospinal fluid system. J Neurochem 69(2):443–454
9. Burdo JR, Antonetti DA, Wolpert EB, Connor JR (2003) Mechanisms and regulation of transferrin and iron transport in a model blood-brain barrier system. Neuroscience 121(4):883–890
10. Burkhart A, Skjorringe T, Johnsen KB, Siupka P, Thomsen LB, Nielsen MS, Thomsen LL, Moos T (2016) Expression of iron-related proteins at the neurovascular unit supports reduction and reoxidation of iron for transport through the blood-brain barrier. Mol Neurobiol 53(10):7237–7253
11. Chen JH, Shahnavas S, Singh N, Ong WY, Walczyk T (2013) Stable iron isotope tracing reveals significant brain iron uptake in adult rats. Metallomics Integr Biometal Sci 5(2):167–173
12. Cooperman SS, Meyron-Holtz EG, Olivierre-Wilson H, Ghosh MC, McConnell JP, Rouault TA (2005) Microcytic anemia, erythropoietic protoporphyria, and neurodegeneration in mice with targeted deletion of iron-regulatory protein 2. Blood 106(3):1084–1091
13. De Domenico I, Ward DM, di Patti MC, Jeong SY, David S, Musci G, Kaplan J (2007) Ferroxidase activity is required for the stability of cell surface ferroportin in cells expressing GPI-ceruloplasmin. EMBO J 26(12):2823–2831
14. De Domenico I, Ward DM, Langelier C, Vaughn MB, Nemeth E, Sundquist WI, Ganz T, Musci G, Kaplan J (2007) The molecular mechanism of hepcidin-mediated ferroportin down-regulation. Mol Biol Cell 18(7):2569–2578
15. Donovan A, Brownlie A, Zhou Y, Shepard J, Pratt SJ, Moynihan J, Paw BH, Drejer A, Barut B, Zapata A, Law TC, Brugnara C, Lux SE, Pinkus GS, Pinkus JL, Kingsley PD, Palis J, Fleming MD, Andrews NC, Zon LI (2000) Positional cloning of zebrafish ferroportin1 identifies a conserved vertebrate iron exporter. Nature 403(6771):776–781
16. Drakesmith H, Nemeth E, Ganz T (2015) Ironing out ferroportin. Cell Metab 22(5):777–787
17. Drayer B, Burger P, Darwin R, Riederer S, Herfkens R, Johnson GA (1986) MRI of brain iron. AJR Am J Roentgenol 147(1):103–110
18. Du F, Qian C, Qian ZM, Wu XM, Xie H, Yung WH, Ke Y (2011) Hepcidin directly inhibits transferrin receptor 1 expression in astrocytes via a cyclic AMP-protein kinase a pathway. Glia 59(6):936–945
19. Du F, Qian ZM, Luo Q, Yung WH, Ke Y (2015) Hepcidin suppresses brain iron accumulation by downregulating iron transport proteins in iron-overloaded rats. Mol Neurobiol 52(1):101–114
20. Eisenstein RS (2000) Iron regulatory proteins and the molecular control of mammalian iron metabolism. Annu Rev Nutr 20:627–662

21. Faucheux BA, Nillesse N, Damier P, Spik G, Mouatt-Prigent A, Pierce A, Leveugle B, Kubis N, Hauw JJ, Agid Y et al (1995) Expression of lactoferrin receptors is increased in the mesencephalon of patients with Parkinson disease. Proc Natl Acad Sci USA 92(21):9603–9607
22. Fillebeen C, Dehouck B, Benaissa M, Dhennin-Duthille I, Cecchelli R, Pierce A (1999) Tumor necrosis factor-alpha increases lactoferrin transcytosis through the blood-brain barrier. J Neurochem 73(6):2491–2500
23. Fillebeen C, Ruchoux MM, Mitchell V, Vincent S, Benaissa M, Pierce A (2001) Lactoferrin is synthesized by activated microglia in the human substantia nigra and its synthesis by the human microglial CHME cell line is upregulated by tumor necrosis factor alpha or 1-methyl-4-phenylpyridinium treatment. Brain Res Mol Brain Res 96(1–2):103–113
24. Fisher J, Devraj K, Ingram J, Slagle-Webb B, Madhankumar AB, Liu X, Klinger M, Simpson IA, Connor JR (2007) Ferritin: a novel mechanism for delivery of iron to the brain and other organs. Am J Physiol Cell Physiol 293(2):C641–C649
25. Galy B, Ferring D, Benesova M, Benes V, Hentze MW (2004) Targeted mutagenesis of the murine IRP1 and IRP2 genes reveals context-dependent RNA processing differences in vivo. RNA (New York, NY) 10(7):1019–1025
26. Ganz T, Nemeth E (2012) Hepcidin and iron homeostasis. Biochimica et biophysica acta 1823(9):1434–1443
27. Gao G, You L-H, Chang Y-Z (2017) Iron metabolism in Parkinson's disease. In: Oxidative stress and redox signalling in Parkinson's disease. The Royal Society of Chemistry, pp 255–276 (Chapter 9)
28. Ghosh MC, Zhang DL, Rouault TA (2015) Iron misregulation and neurodegenerative disease in mouse models that lack iron regulatory proteins. Neurobiol Dis 81:66–75
29. Gunshin H, Mackenzie B, Berger UV, Gunshin Y, Romero MF, Boron WF, Nussberger S, Gollan JL, Hediger MA (1997) Cloning and characterization of a mammalian proton-coupled metal-ion transporter. Nature 388(6641):482–488
30. Han J, Day JR, Connor JR, Beard JL (2003) Gene expression of transferrin and transferrin receptor in brains of control vs. iron-deficient rats. Nutr Neurosci 6(1):1–10
31. Healy S, McMahon J, Owens P, FitzGerald U (2016) Significant glial alterations in response to iron loading in a novel organotypic hippocampal slice culture model. Sci Rep 6:36410
32. Healy S, McMahon JM, FitzGerald U (2017) Modelling iron mismanagement in neurodegenerative disease in vitro: paradigms, pitfalls, possibilities & practical considerations. Prog Neurobiol 158:1–14
33. Hentze MW, Muckenthaler MU, Andrews NC (2004) Balancing acts: molecular control of mammalian iron metabolism. Cell 117(3):285–297
34. Hentze MW, Muckenthaler MU, Galy B, Camaschella C (2010) Two to tango: regulation of Mammalian iron metabolism. Cell 142(1):24–38
35. Jeong SY, David S (2003) Glycosylphosphatidylinositol-anchored ceruloplasmin is required for iron efflux from cells in the central nervous system. J Biol Chem 278(29):27144–27148
36. Kautz L, Jung G, Du X, Gabayan V, Chapman J, Nasoff M, Nemeth E, Ganz T (2015) Erythroferrone contributes to hepcidin suppression and iron overload in a mouse model of beta-thalassemia. Blood 126(17):2031–2037
37. Kautz L, Jung G, Valore EV, Rivella S, Nemeth E, Ganz T (2014) Identification of erythroferrone as an erythroid regulator of iron metabolism. Nat Genet 46(7):678–684
38. Ke Y, Qian ZM (2003) Iron misregulation in the brain: a primary cause of neurodegenerative disorders. Lancet Neurol 2(4):246–253
39. Ke Y, Qian ZM (2007) Brain iron metabolism: neurobiology and neurochemistry. Prog Neurobiol 83(3):149–173
40. Kim HY, LaVaute T, Iwai K, Klausner RD, Rouault TA (1996) Identification of a conserved and functional iron-responsive element in the 5′-untranslated region of mammalian mitochondrial aconitase. J Biol Chem 271(39):24226–24230
41. Klomp LW, Farhangrazi ZS, Dugan LL, Gitlin JD (1996) Ceruloplasmin gene expression in the murine central nervous system. J Clin Investig 98(1):207–215

42. Klomp LW, Gitlin JD (1996) Expression of the ceruloplasmin gene in the human retina and brain: implications for a pathogenic model in aceruloplasminemia. Hum Mol Genet 5(12):1989–1996
43. Krause A, Neitz S, Magert HJ, Schulz A, Forssmann WG, Schulz-Knappe P, Adermann K (2000) LEAP-1, a novel highly disulfide-bonded human peptide, exhibits antimicrobial activity. FEBS Lett 480(2–3):147–150
44. Lai CH, Kuo KH (2005) The critical component to establish in vitro BBB model: Pericyte. Brain Res Brain Res Rev 50(2):258–265
45. Lane DJ, Robinson SR, Czerwinska H, Bishop GM, Lawen A (2010) Two routes of iron accumulation in astrocytes: ascorbate-dependent ferrous iron uptake via the divalent metal transporter (DMT1) plus an independent route for ferric iron. Biochem J 432(1):123–132
46. LaVaute T, Smith S, Cooperman S, Iwai K, Land W, Meyron-Holtz E, Drake SK, Miller G, Abu-Asab M, Tsokos M, Switzer R 3rd, Grinberg A, Love P, Tresser N, Rouault TA (2001) Targeted deletion of the gene encoding iron regulatory protein-2 causes misregulation of iron metabolism and neurodegenerative disease in mice. Nat Genet 27(2):209–214
47. Li H, Sun H, Qian ZM (2002) The role of the transferrin-transferrin-receptor system in drug delivery and targeting. Trends Pharmacol Sci 23(5):206–209
48. Ma J, Zhang FL, Zhou G, Bao YX, Shen Y, Qian ZM (2018) Different characteristics of hepcidin expression in IL-6+/+ and IL-6-/- neurons and astrocytes treated with lipopolysaccharides. Neurochem Res 43(8):1624–1630
49. Malecki EA, Devenyi AG, Beard JL, Connor JR (1999) Existing and emerging mechanisms for transport of iron and manganese to the brain. J Neurosci Res 56(2):113–122
50. McCarthy RC, Kosman DJ (2013) Ferroportin and exocytoplasmic ferroxidase activity are required for brain microvascular endothelial cell iron efflux. J Biol Chem 288(24):17932–17940
51. McCarthy RC, Kosman DJ (2014) Glial cell ceruloplasmin and hepcidin differentially regulate iron efflux from brain microvascular endothelial cells. PLoS ONE 9(2):e89003
52. McCarthy RC, Kosman DJ (2015) Mechanisms and regulation of iron trafficking across the capillary endothelial cells of the blood-brain barrier. Front Mol Neurosci 8:31
53. McKie AT, Barrow D, Latunde-Dada GO, Rolfs A, Sager G, Mudaly E, Mudaly M, Richardson C, Barlow D, Bomford A, Peters TJ, Raja KB, Shirali S, Hediger MA, Farzaneh F, Simpson RJ (2001) An iron-regulated ferric reductase associated with the absorption of dietary iron. Science (New York, NY) 291(5509):1755–1759
54. Meyron-Holtz EG, Ghosh MC, Iwai K, LaVaute T, Brazzolotto X, Berger UV, Land W, Ollivierre-Wilson H, Grinberg A, Love P, Rouault TA (2004) Genetic ablations of iron regulatory proteins 1 and 2 reveal why iron regulatory protein 2 dominates iron homeostasis. EMBO J 23(2):386–395
55. Moos T (2002) Brain iron homeostasis. Dan Med Bull 49(4):279–301
56. Moos T, Morgan EH (2000) Transferrin and transferrin receptor function in brain barrier systems. Cell Mol Neurobiol 20(1):77–95
57. Moos T, Rosengren Nielsen T, Skjorringe T, Morgan EH (2007) Iron trafficking inside the brain. J Neurochem 103(5):1730–1740
58. Moos T, Trinder D, Morgan EH (2000) Cellular distribution of ferric iron, ferritin, transferrin and divalent metal transporter 1 (DMT1) in substantia nigra and basal ganglia of normal and beta2-microglobulin deficient mouse brain. Cell Mol Biol (Noisy-le-Grand, France) 46(3):549–561
59. Moroo I, Ujiie M, Walker BL, Tiong JW, Vitalis TZ, Karkan D, Gabathuler R, Moise AR, Jefferies WA (2003) Identification of a novel route of iron transcytosis across the mammalian blood-brain barrier. Microcirculation (New York, NY: 1994) 10(6):457–462
60. Morris CM, Candy JM, Oakley AE, Bloxham CA, Edwardson JA (1992) Histochemical distribution of non-haem iron in the human brain. Acta Anat 144(3):235–257
61. Muckenthaler MU, Galy B, Hentze MW (2008) Systemic iron homeostasis and the iron-responsive element/iron-regulatory protein (IRE/IRP) regulatory network. Annu Rev Nutr 28:197–213

62. Nemeth E, Ganz T (2006) Regulation of iron metabolism by hepcidin. Annu Rev Nutr 26:323–342
63. Nemeth E, Tuttle MS, Powelson J, Vaughn MB, Donovan A, Ward DM, Ganz T, Kaplan J (2004) Hepcidin regulates cellular iron efflux by binding to ferroportin and inducing its internalization. Science (New York, NY) 306(5704):2090–2093
64. Oshiro S, Kawahara M, Kuroda Y, Zhang C, Cai Y, Kitajima S, Shirao M (2000) Glial cells contribute more to iron and aluminum accumulation but are more resistant to oxidative stress than neuronal cells. Biochim Biophys Acta 1502(3):405–414
65. Oshiro S, Morioka MS, Kikuchi M (2011) Dysregulation of iron metabolism in Alzheimer's disease, Parkinson's disease, and amyotrophic lateral sclerosis. Adv Pharmacol Sci 2011:378278
66. Pantopoulos K (2004) Iron metabolism and the IRE/IRP regulatory system: an update. Ann N Y Acad Sci 1012:1–13
67. Patel BN, David S (1997) A novel glycosylphosphatidylinositol-anchored form of ceruloplasmin is expressed by mammalian astrocytes. J Biol Chem 272(32):20185–20190
68. Patel BN, Dunn RJ, David S (2000) Alternative RNA splicing generates a glycosylphosphatidylinositol-anchored form of ceruloplasmin in mammalian brain. J Biol Chem 275(6):4305–4310
69. Pelizzoni I, Zacchetti D, Campanella A, Grohovaz F, Codazzi F (2013) Iron uptake in quiescent and inflammation-activated astrocytes: a potentially neuroprotective control of iron burden. Biochim Biophys Acta 1832(8):1326–1333
70. Peters DG, Connor JR, Meadowcroft MD (2015) The relationship between iron dyshomeostasis and amyloidogenesis in Alzheimer's disease: two sides of the same coin. Neurobiol Dis 81:49–65
71. Qian ZM, Chang YZ, Leung G, Du JR, Zhu L, Wang Q, Niu L, Xu YJ, Yang L, Ho KP, Ke Y (2007) Expression of ferroportin1, hephaestin and ceruloplasmin in rat heart. Biochem Biophys Acta 1772(5):527–532
72. Qian ZM, He X, Liang T, Wu KC, Yan YC, Lu LN, Yang G, Luo QQ, Yung WH, Ke Y (2014) Lipopolysaccharides upregulate hepcidin in neuron via microglia and the IL-6/STAT3 signaling pathway. Mol Neurobiol 50(3):811–820
73. Qian ZM, Pu YM, Wang Q, Ke Y, Yao YD, Chen WF, Shen X, Feng YM, Tang PL (1999) Cerebellar granule cells acquire transferrin-free iron by a carrier-mediated process. Neuroscience 92(2):577–582
74. Qian ZM, Tang PL, Wang Q (1997) Iron crosses the endosomal membrane by a carrier-mediated process. Prog Biophys Mol Biol 67(1):1–15
75. Ramey G, Deschemin JC, Durel B, Canonne-Hergaux F, Nicolas G, Vaulont S (2010) Hepcidin targets ferroportin for degradation in hepatocytes. Haematologica 95(3):501–504
76. Rasmussen MK, Mestre H, Nedergaard M (2018) The glymphatic pathway in neurological disorders. Lancet Neurol 17(11):1016–1024
77. Rouault TA (2013) Iron metabolism in the CNS: implications for neurodegenerative diseases. Nat Rev Neurosci 14(8):551–564
78. Rouault TA, Zhang DL, Jeong SY (2009) Brain iron homeostasis, the choroid plexus, and localization of iron transport proteins. Metab Brain Dis 24(4):673–684
79. Ryan F, Zarruk JG, Losslein L, David S (2018) Ceruloplasmin plays a neuroprotective role in cerebral ischemia. Frontiers in neuroscience 12:988
80. Siddappa AJ, Rao RB, Wobken JD, Leibold EA, Connor JR, Georgieff MK (2002) Developmental changes in the expression of iron regulatory proteins and iron transport proteins in the perinatal rat brain. J Neurosci Res 68(6):761–775
81. Skjorringe T, Moller LB, Moos T (2012) Impairment of interrelated iron- and copper homeostatic mechanisms in brain contributes to the pathogenesis of neurodegenerative disorders. Front Pharmacol 3:169
82. Smith SR, Ghosh MC, Ollivierre-Wilson H, Hang Tong W, Rouault TA (2006) Complete loss of iron regulatory proteins 1 and 2 prevents viability of murine zygotes beyond the blastocyst stage of embryonic development. Blood Cells Mol Dis 36(2):283–287

83. Song N, Wang J, Jiang H, Xie J (2018) Astroglial and microglial contributions to iron metabolism disturbance in Parkinson's disease. Biochimica et biophysica acta 1864(3):967–973
84. Urrutia P, Aguirre P, Esparza A, Tapia V, Mena NP, Arredondo M, Gonzalez-Billault C, Nunez MT (2013) Inflammation alters the expression of DMT1, FPN1 and hepcidin, and it causes iron accumulation in central nervous system cells. J Neurochem 126(4):541–549
85. Vargas JD, Herpers B, McKie AT, Gledhill S, McDonnell J, van den Heuvel M, Davies KE, Ponting CP (2003) Stromal cell-derived receptor 2 and cytochrome b561 are functional ferric reductases. Biochem Biophys Acta 1651(1–2):116–123
86. Wang J, Jiang H, Xie JX (2007) Ferroportin1 and hephaestin are involved in the nigral iron accumulation of 6-OHDA-lesioned rats. Eur J Neurosci 25(9):2766–2772
87. Wang L, Liu X, You LH, Ci YZ, Chang S, Yu P, Gao G, Chang YZ (2019) Hepcidin and iron regulatory proteins coordinately regulate ferroportin 1 expression in the brain of mice. J Cell Physiol 234(5):7600–7607
88. Wang SM, Fu LJ, Duan XL, Crooks DR, Yu P, Qian ZM, Di XJ, Li J, Rouault TA, Chang YZ (2010) Role of hepcidin in murine brain iron metabolism. Cell Mol Life Sci CMLS 67(1):123–133
89. Ward RJ, Zucca FA, Duyn JH, Crichton RR, Zecca L (2014) The role of iron in brain ageing and neurodegenerative disorders. Lancet Neurol 13(10):1045–1060
90. Wu LJ, Leenders AG, Cooperman S, Meyron-Holtz E, Smith S, Land W, Tsai RY, Berger UV, Sheng ZH, Rouault TA (2004) Expression of the iron transporter ferroportin in synaptic vesicles and the blood-brain barrier. Brain Res 1001(1–2):108–117
91. Xu H, Wang Y, Song N, Wang J, Jiang H, Xie J (2017) New progress on the role of glia in iron metabolism and iron-induced degeneration of dopamine neurons in Parkinson's disease. Frontiers in molecular neuroscience 10:455
92. Yang WM, Jung KJ, Lee MO, Lee YS, Lee YH, Nakagawa S, Niwa M, Cho SS, Kim DW (2011) Transient expression of iron transport proteins in the capillary of the developing rat brain. Cell Mol Neurobiol 31(1):93–99
93. Yu P, Qian ZM, Duan XL, Chang YZ (2004) The structure, gene expression of Divalent Metal Transporter 1 and the underlying regulatory mechanisms. Chin J Neuroanat 20:205–209
94. Zanardi A, Conti A, Cremonesi M, D'Adamo P, Gilberti E, Apostoli P, Cannistraci CV, Piperno A, David S, Alessio M (2018) Ceruloplasmin replacement therapy ameliorates neurological symptoms in a preclinical model of aceruloplasminemia. EMBO Mol Med 10(1):91–106
95. Zechel S, Huber-Wittmer K, von Bohlen und Halbach O (2006) Distribution of the iron-regulating protein hepcidin in the murine central nervous system. J Neurosci Res 84(4):790–800
96. Zhang P, Land W, Lee S, Juliani J, Lefman J, Smith SR, Germain D, Kessel M, Leapman R, Rouault TA, Subramaniam S (2005) Electron tomography of degenerating neurons in mice with abnormal regulation of iron metabolism. J Struct Biol 150(2):144–153
97. Zheng W, Monnot AD (2012) Regulation of brain iron and copper homeostasis by brain barrier systems: implication in neurodegenerative diseases. Pharmacol Ther 133(2):177–188
98. Zumbrennen-Bullough KB, Becker L, Garrett L, Holter SM, Calzada-Wack J, Mossbrugger I, Quintanilla-Fend L, Racz I, Rathkolb B, Klopstock T, Wurst W, Zimmer A, Wolf E, Fuchs H, Gailus-Durner V, de Angelis MH, Romney SJ, Leibold EA (2014) Abnormal brain iron metabolism in Irp2 deficient mice is associated with mild neurological and behavioral impairments. PLoS ONE 9(6):e98072

Chapter 4
Iron Pathophysiology in Parkinson Diseases

Hong Jiang, Ning Song, Qian Jiao, Limin Shi and Xixun Du

Abstract The key molecular events that provoke Parkinson's disease (PD) are not fully understood. Iron deposit was found in the substantia nigra pars compacta (SNpc) of PD patients and animal models, where dopaminergic neurons degeneration occurred selectively. The mechanisms involved in disturbed iron metabolism remain unknown, however, considerable evidence indicates that iron transporters dysregulation, activation of L-type voltage-gated calcium channel (LTCC) and ATP-sensitive potassium (KATP) channels, as well as N-methyl-D-aspartate (NMDA) receptors (NMDARs) contribute to this process. There is emerging evidence on the structural links and functional modulations between iron and α-synuclein, and the key player in PD which aggregates in Lewy bodies. Iron is believed to modulate α-synuclein synthesis, post-translational modification, and aggregation. Furthermore, glia, especially activated astroglia and microglia, are involved in iron deposit in PD. Glial contributions were largely dependent on the factors they released, e.g., neurotrophic factors, pro-inflammatory factors, lactoferrin, and those undetermined. Therefore, iron chelation using iron chelators, the extracts from many natural foods with iron chelating properties, may be an effective therapy for prevention and treatment of the disease.

Keywords Iron · Parkinson's disease · α-Synuclein · Glia · Iron chelation

4.1 Iron Metabolism Dysfunction in Parkinson's Diseases

Parkinson's disease (PD), the second common neurodegenerative disorder of the elderly, is characterized by the death of dopaminergic neurons in the substantia nigra (SN) pars compacta (SNpc), leading to striatal dopamine (DA) deficiency [1–3]. PD affects 1–2% of those over 60 years of age by causing motor dysfunctions such as bradykinesia, resting tremor, rigidity, postural instability, as well as non-motor symptoms, such as cognition, hyposmia, sleep disturbances, depression, constipa-

H. Jiang (✉) · N. Song · Q. Jiao · L. Shi · X. Du
Department of Physiology, Medical College of Qingdao University, Qingdao 266071, China
e-mail: hongjiang@qdu.edu.cn

Y.-Z. Chang (ed.), *Brain Iron Metabolism and CNS Diseases*,
Advances in Experimental Medicine and Biology 1173,
https://doi.org/10.1007/978-981-13-9589-5_4

tion, and other dysautonomic symptoms [4–6]. Up to now, the key molecular events that provoke neurodegeneration are not fully understood. In addition to misfolding of proteins, impairment of the ubiquitin proteasome, dysfunction of mitochondria, oxidative stress, there is mounting evidence implicated that iron accumulation is pivotal to PD pathogenesis [7, 8].

In the late 1980s, several groups reported elevated iron levels in the SN of parkinsonian patients compared to age-matched control [9–12]. In other parts of brain, including cortex (Brodmann area 21), hippocampus, putamen, and globus pallidus, the contents of iron were unchanged [9]. Especially, it was reported that SNpc is a selective area of iron accumulation and the ratio of ferrous to ferric iron decreased in parkinsonian patients [13]. Furthermore, a variety of methods, such as inductively coupled plasma spectroscopy, magnetic resonance imaging, laser microprobe mass analysis, susceptibility-weighted imaging, and enhanced T2-star-weighted angiography, have proved the increased iron levels in the SNpc of PD brains [7]. The disruption of iron metabolism in parkinsonian patients is also demonstrated by a previous work that the iron levels are significantly decreased in the temporal cortex in brain when compared with age-matched healthy controls, suggesting a likely re-distribution pattern in PD [14].

The iron metabolism disturbance was also present in PD animal and cell models. The transgenic PD mice model overexpressing the mutant human A53T α-synuclein (A53T mice) show an accumulation of aggregates comprising α-synuclein and develop age-dependent motor deficits. A53T mice are more vulnerable to accumulate iron in the SN than the wild type [15]. 1-methyl-4-phenyl-1,2,3,6-tetrahydropyridine (MPTP), an inhibitor of complex I of the mitochondrial electron transport chain, is commonly used to induce PD animal models that exhibit the damage of dopaminergic neurons [1, 16]. Injection of MPTP into the caudate or putamen of monkeys resulted in the increase of ferric iron-reaction products in the ipsilateral SNpc [17, 18]. In hemiparkinsonian monkeys, iron contents were significantly increased from 4.5 to 18 months in the ipsilateral SNpc after MPTP injection by unilateral internal carotid artery [19]. Moreover, the increasing extent was significantly correlated to the extent of dopaminergic cell death [17, 19, 20]. Additionally, the phenomena that nigral iron levels elevation is also observed in MPTP-induced PD mice model [21–23]. 6-hydroxydopamine (6-OHDA) is a neurotoxin, which induces dopaminergic neurons death and therefore has been implicated in PD rat models. It was reported that iron levels increased in the ipsilateral SN from 1 to 14 days in 6-OHDA unilaterally lesioned rats [24]. Actually, both increased iron levels and dopaminergic neurons loss are apparent in the SN as promptly as one day following 6-OHDA injection [24]. Iron levels increased in SN with 6-OHDA injection into rats have been proved by peer laboratories [21, 25–27]. In vitro studies, iron uptake significantly increased both in 1-methyl-4-phenylpyridinium (MPP^+)- and 6-OHDA-induced PD cell models [28, 29]. In other models of PD, including lipopolysaccharide-induced inflammation models, the loss of dopaminergic neurons was accompanied by increased iron and ferritin levels in glial cells of the SN pars reticulate [30, 31]. Moreover, in iron-overloaded animal models, such as direct injection of ferric iron into the SN, high dietary iron supplements and peripheral iron overload, they develop degeneration of dopaminergic neurons and further illustrated the relationship between iron and PD pathogenesis [15, 32]. With the common focusing on neurotoxins MPTP and 6-OHDA, the environment toxins exiting in food and drinking water, including paraquat, rotenone, and mancozeb, may be directly or indirectly involved and affect iron metabolism [33, 34].

The unbalanced distribution of iron in the brain of parkinsonian patients indicates that iron plays a key role in the degeneration of dopaminergic neurons. Potential mechanisms have been offered to explain why iron metabolism disturbance occurs (see below), as well as how elevated iron behaviors leading to dopaminergic neurons degeneration [7]. Among these, iron and DA toxic couple has to be highlighted. DA is the central neurotransmitter involved in PD. DA easily forms toxic metabolites and these processes occur predominantly in the cytoplasm [35]. Physiologically, H_2O_2 is produced by a DA enzymatic process via monoamine oxidase. The revealed aspects of iron-DA coupling are largely derived from their pro-oxidant properties [36]. The high-level iron in dopaminergic neurons aggravates the oxidative stress by Fenton reaction that iron reacts with H_2O_2 produced in the enzymatic process in DA metabolism and then can generate OH that can damage proteins, nucleic acids, and membrane phospholipids [37]. An alternative mechanism has been raised that the oxidation of DA by iron forms 6-OHDA. As a commonly used neurotoxin in PD models, 6-OHDA even liberates iron from ferritin and produced H_2O_2 in its metabolism. Recently, it was reported that high-level iron inside cells caused ferroptosis which is a form of regulated cell death characterized by the iron-dependent accumulation of lipid hydroperoxides to lethal levels [38, 39]. Elevated iron likely contributes in dopaminergic neurons degeneration in PD, which could be further validated by the facts that using pharmacological or genetic chelation of iron in animal models of PD has neuroprotective effects on dopaminergic neurons (see below) [40, 41].

4.2 Possible Mechanisms Underlying Iron Metabolism Disturbance

4.2.1 Roles of Iron Transporters

The mechanisms involved in this iron accumulation remain unknown. Considerable evidence indicates that dysregulation of several iron transporters (responsible for either iron influx/uptake or iron efflux/release) contributes to the abnormal iron deposit. Thus, understanding iron transporters dysregulation, particularly in the context of PD, might be beneficial for exploring therapeutic strategies via restoring iron homeostasis.

Generally, cellular iron uptake from transferrin (Tf) is accomplished by glycoprotein-receptor-mediated endocytosis via transferrin receptor (TfR), which is considered as a major pathway for iron import in brain. There are two types of

TfR, namely TfR1 and TfR2. TfR1, expressed abundantly in the central nervous system, especially in neurons, constitutes the main receptor for Tf and internalizes it through receptor-mediated endocytosis. Furthermore, TfR1 mRNA contains iron-responsive elements (IRE) in the 5′-untranslated region, thus it can be regulated by intracellular iron levels via iron response proteins (IRP) [42]. In contrast, TfR2, mainly expressed in the mitochondria of nigral dopaminergic neurons [43], does not contain the IRE, thus it cannot be regulated by iron levels. Although TfR2 has a lower affinity for Tf than TfR1, it can serve as an iron sensor in the regulation of hepcidin [44]. Recent studies found that 1-methyl-4-phenylpyridinium (MPP^+) treatment led to the increased expression of TfR, suggesting a role for TfR-dependent iron influx in the selective iron accumulation in PD [45]. The localization of TfR2 in the mitochondria of dopaminergic neurons and a dramatic increase of oxidized Tf in the SN raised the possibility that the Tf-TfR2 system might also contribute to iron deposit in PD [46].

The non-transferrin-bound iron (NTBI) is mainly transported through divalent metal transporter 1 (DMT1), which is responsible for ferrous iron uptake. There are at least four distinct DMT1 mRNA isoforms: 1A, 1B, +IRE and −IRE [47, 48]. DMT1, also known as natural resistance-associated macrophage protein 2 (Nramp 2) or divalent cation transporter 1 (DCT1), was first discovered to transport iron in the duodenal lumen. In the SN, DMT1 is widely expressed in neurons, astrocytes and microglia [49], which mediates iron transmembrane transport as well as from the endosomes to the cytoplasm [50]. The expression of DMT1 can be regulated post-transcriptionally through IRE/IRP, or post-translationally through the ubiquitin-proteasome system (UPS). Previous studies have shown that the E3 ligase parkin is responsible for the proteasomal degradation of DMT1 and regulates iron transport [51]. Moreover, DMT1 can be ubiquitinated by the neural precursor cell expressed developmentally down-regulated protein 4 (Nedd4) family member WW domain-containing protein 2 (WWP2), and this requires Nedd4 family interacting protein 1 (Ndfip1) and/or Ndfip2 [52]. In the postmortem brains of PD patients and animal models, the expression of nigral DMT1 was significantly up-regulated, thus contributed to the intracellular iron accumulation and dopaminergic neurons degeneration [28, 53, 54]. Modulation of DMT1 proteasomal degradation via UPS might also be associated with DMT1 dysregulation and participate in iron deposit in PD [51, 55, 56].

Another iron uptake mechanism involves an iron transporter called lactoferrin (Lf), which acts as an iron-binding protein, belonging to the Tf family. In a similar manner to Tf, Lf-bound ferric iron can bind to Lf receptor (LfR), resulting in iron transport across the plasma membrane. The affinity of Lf for iron is 300 times higher than that of Tf [57]. In the brain, Lf is only synthesized and released by activated microglia [58]. LfR is present in blood vessels and nigral dopaminergic neurons [59], suggesting that Lf-LfR pathway may be involved in iron transport into brain parenchyma and dopaminergic neurons. Moreover, an increased expression of Lf and LfR on dopaminergic neurons was found in PD patients [58, 59], which may account for the excessive accumulation of iron. And this increase was occurred in the most severely affected dopaminergic neurons, indicating a relationship between Lf-LfR increase and dopaminergic neurons degeneration.

Up to now, ferroportin 1 (FPN1) is the only known iron transporter responsible for intracellular iron export. Iron is exported in the form of ferrous iron, therefore the ferroxidase activity of hephaestin (HP) or ceruloplasmin (CP) is needed, which oxidizes ferrous iron (Fe^{2+}) into the ferric form (Fe^{3+}). FPN1 and HP co-localized in neurons, astrocytes, oligodendrocytes, and microglia, indicating that HP could facilitate FPN1-mediated iron efflux [60]. CP acts as a copper-dependent ferroxidase, which can exist as a soluble form in plasma or as a membrane-bound form highly expressed in astrocytes. The absence of CP in aceruloplasminemia results in iron accumulation in the basal ganglia accompanied by neuronal degeneration [61]. It has been also reported that iron exporter FPN1 was reduced in the brain of 6-OHDA/MPTP/lipopolysaccharide-treated animal models [31, 60, 62]. In addition, several evidence implicates the role of CP in the pathogenesis of PD. Reduced ferroxidase activity of CP was found in the CSF of PD patients [63]. The decreased expression of CP in the SN is involved in the nigral iron accumulation in 6-OHDA-lesioned rats [64].

4.2.2 Roles of L-Type Voltage-Gated Calcium Channel (LTCC)

The LTCC of SN dopaminergic neurons contains pore-forming Cav1.2 or Cav1.3 subunits in complex with an accessory β and α2-δ subunit [65]. Compared with ventral tegmental area (VTA) dopaminergic neurons, LTCC are selectively activated in the nigral dopaminergic neurons during pacemaking and cause an oscillating calcium burden inducing mitochondrial stress. Otherwise, the δ subunits of LTCC were obviously up-regulated in the proteome analysis of human SN in PD. These studies suggest that LTCC might be related to the pathogenesis of PD [66]. In cardiomyocytes, secretory cells and neurons, LTCC have the similar functional properties, which could mediate iron import into cells which were iron overload [67, 68]. Some evidence has demonstrated that iron could compete with calcium via LTCC for entry into NGF-treated rat PC12 cells and murine N2a cells, which suggest that LTCC could provide an alternative route for iron entering into neuronal cells under the iron-overloaded conditions [69]. In the SN of rats, the protective effect of LTCC blocker nifedipine against iron overload-induced dopaminergic neurons degeneration and iron accumulation has been demonstrated [70]. Based on these studies, LTCC might mediate the iron overload in the dopaminergic neurons and be involved in the selective iron accumulation in the SN in PD.

4.2.3 Roles of N-Methyl-D-Aspartate (NMDA) Receptors (NMDARs)

NMDARs are cation channels which mediate Na^+ and Ca^{2+} ions entering the cells and could be activated by glutamate, glycine, or D-serineare. NMDARs activation appropriately plays a critical role in several physiological processes, e.g., synaptic plasticity and excitatory neurotransmission. However, excessive activation of NMDARs contributes to pathological changes in the brain. NMDARs mediated excessive Ca^{2+} influx might induce neuronal death in several neurodegenerative diseases, including PD [71, 72]. NR1, a very important subunit of NMDARs, which expression was increased in surviving dopaminergic neurons from PD brains compared with the controls [73]. Some studies also revealed that NMDARs activation could significantly promote Fe^{2+} entry into cells. The underlying mechanisms might be involved. First, NMDARs activation increases NTBI influx. Fe^{2+} could compete with Ca^{2+} for NMDARs to enter the cultured neurons [74, 75]. The increased iron levels may induce a corresponding ROS overproduction and higher susceptibility of neurons. Another investigation has identified a novel signaling cascade for NMDARs regulated iron uptake in brain [76]. It was reported that glutamate via NMDARs trigger Ca^{2+} influx, then activated neuronal nitric oxide (NO) synthase (NOS) produce NO, which induce excitability toxicity [76–78]. And neuronal NOS (nNOS) binding to CAPON could cause NO delivery to Dexrase1 and S-nitrosylation of Dexrase1, which promote iron uptake by regulating the function of DMT1 [79]. Rhes (Dexras2), a homolog of Dexras1, is selectively localized to the corpus striatum. Rhes was also reported to be involved in NMDARs-induced iron influx into the striatal neurons and regulating striatal iron homeostasis by PKA-Rhes-DMT1 pathway [80].

Second, NMDARs activation enhanced iron releasing from lysosome. The iron in lysosome serves as a main source for intracellular iron, in which DMT1 plays a critical role in iron recycling from lysosome to cytoplasm. Recently, several evidence showed that Dexras1/PAP7/DMT1 complex located on the lysosome. Either impaired DMT1 function or the collapsed proton gradient could reduce iron pool in cytoplasm, which indicated that Dexras1/PAP7/DMT1 complex plays a role in iron release from lysosome to cytoplasm [81].

Finally, NMDARs activation increased DMT1 expression. It has been reported that the mRNA levels of DMT1 were increased after 5 min treated with 50 μM NMDA to primary hippocampal cultures. With exposure to the transcription inhibitor actinomycin D and NMDARs antagonist MK-801, DMT1 up-regulation induced by NMDAs or 6-OHDA could be restored [82]. Interestingly, both excessive activation of NMDAs and up-regulation of DMT1 were observed in the process of degeneration of dopaminergic neurons in PD, indicating the close links between NMDARs and DMT1 in PD [83].

4.2.4 Roles of ATP-Sensitive Potassium (K_{ATP}) Channels

Functional K_{ATP} channels are composed of four pore-forming Kir6 (Kir6.1 and Kir6.2) subunits and four regulatory SUR (SUR1, SUR2A, and SUR2B) subunits. Different subunit compositions decided the different biophysical, pharmacological, and metabolic properties of K_{ATP} channels [84]. In dopaminergic neurons in the SN and VTA, K_{ATP} channels are both expressed with Kir6.2 and SUR1 subunits. But only the K_{ATP} channels of dopaminergic neurons in the SN were selectively activated when treated with MPP^+. And dopaminergic neurons in the SN but not those in the VTA in the MPTP model were selectively rescued after genetic inactivation of Kir6.2 subunits [85]. It was also reported that mRNA levels of the regulatory subunit SUR1 were increased in survived dopaminergic neurons in the SN from PD patients. Otherwise, mRNA levels of SUR2 and Kir6.2 were not changed [73]. These results suggest that the activation of K_{ATP} channels might be involved in the selective degeneration of dopaminergic neurons in the SN in PD. The activation of K_{ATP} channels would induce the membrane potential hyperpolarized, which is important for the membrane oscillations that underlie bursting firing generated [86]. K_{ATP} channels' gated burst firing could aggravate excitotoxicity and increase calcium loading with NMDARs and LTCCs synergistically in the dopaminergic neurons of SN, where metabolic state have already challenged [87]. This led to the insufficient mitochondrial calcium buffering capacities and acceleration of calcium-induced ROS production [88], which in turn activated K_{ATP} channels in highly vulnerable neurons [73].

Selective activation of K_{ATP} channels in the SN was suggested to contribute to iron accumulation. The transport function of DMT1 is proton-coupled and dependent on the cell membrane potential. It was reported that hyperpolarized potentials could promote the iron uptake via DMT1 [89]. There is evidence that diazoxide, a K_{ATP} channel opener, could increase intracellular iron levels after ferrous iron treatment [90]. The subsequent ATP depletion and ROS production would induce additional K_{ATP} channels activated in a feed-forward cycle [73]. Accordingly, inhibition of K_{ATP} channels dramatically decreases the iron uptake and inhibits cell damage [90]. These provide evidence that the selective iron accumulation in the SN might be related to K_{ATP} channels activation; further investigation should be done to reveal the underlying mechanisms.

4.2.5 Iron and Gene Mutation

Parkin, α-synuclein, DJ-1, leucine-rich repeat kinase 2 (LRRK2), PTEN-induced kinase 1 (PINK1), and several other genes were believed to be involved in the pathogenesis of PD. As a matter of fact, these genes were demonstrated to be linked to iron accumulation in PD [51, 91]. PD patients with Parkin, α-synuclein, DJ-1, LRRK2, or PINK1 mutations showed significantly larger echogenicity in the SN compared with the healthy control group evaluated by transcranial sonography [43], which indicated an increased nigral iron level. Thus, hyperechogenicity has been partially shown to be increased iron levels in the SN [92], which has also been deemed to be a typical sign in idiopathic PD. A subset of cellular proteins not degraded attributed to Parkin mutations are currently supposed to be the most common reasons of familial Parkinsonism [93–95]. Iron could induce altered Parkin solubility and result in its intracellular aggregation. And with the depletion of soluble, functional forms of Parkin, proteasomal activities were impaired with cell damage [96]. In 1998, Parkin mutation was first identified in autosomal recessive juvenile parkinsonism (ARJP), and iron staining in the SN was reported to be more intense than that of controls, as well as sporadic PD patients [91]. It was hypothesized that iron accumulation might be related to the loss of the Parkin gene.

More recently, Parkin was reported to be responsible for ubiquitination of DMT1 + IRE. The expression of 1B-DMT1 isoforms was decreased if SH-SY5Y cells overexpressed with Parkin [51]. When fed with iron-supplemented diet, transgenic mice with DMT1 overexpression showed selectively accumulated iron in the SN, meanwhile, the expression of Parkin is also up-regulated, likely a neuroprotective response [97]. In human lymphocytes containing a homozygous deletion of exon 4 of Parkin or in the brains of Parkin knockout animals, expression of DMT1 + IRE was also shown to be elevated.

4.3 Iron Deposit and α-Synuclein Pathology in PD

Among the contributors in PD pathogenesis, α-synuclein was considered to be a key player which aggregates in Lewy bodies (LBs), the neuropathological hallmarks in both familial and sporadic PD. As a coexisting factor with α-synuclein in LB, iron was most pronouncedly stained in the LBs core of the remaining dopaminergic neurons of SNpc [98]. Fe^{3+} has the direct, high affinity for α-synuclein at D121, N122, and E123 [99]. The potential bridging by polyvalent iron is proposed to be a key factor in the metal-induced conformational changes of α-synuclein, leading to significant accelerations in the rate of α-synuclein fibril formation [100]. Fe^{3+} also was able to alter the morphology of α-synuclein fibrils and accelerates aggregation in both wild and mutant α-synuclein, as demonstrated by transmission electron microscopy monitoring α-synuclein aggregation [101]. With the emerging evidence on the structural links and functional modulations between iron and α-synuclein, iron is believed to modulate on α-synuclein synthesis, post-translational modification and aggregation.

It was first predicted that 46 nucleotide in the 5′-UTR of α-synuclein transcript form a single RNA stem-loop, showing similarity to the IREs in the 5′-UTRs of mRNAs encoding H-ferritin, L-ferritin, FPN1, and erythroid 5-aminolevulinate (eALAS) and mitochondrial aconitase, including 100% homology at the five nucleotide apex (CAGUG) [102]. This provides the evidence that iron might be able to regulate α-synuclein synthesis by IRPs-IRE post-transcriptional machinery. In human SK-N-SH cells, α-synuclein mRNA level was up-regulated with IRP1 knockdown, mimicking the condition that IRPs failed to bind IRE with iron [103].

Iron accumulation and oxidative stress induced by ferrous ammonium was able to up-regulate α-synuclein protein levels in these cells. The data showed the 5′-UTR in α-synuclein is a positive mediator of promoting translation, but not of mRNA stability/expression, consistent with the fact that IRPs serve to block transcription activation [104]. On the other hand, in HEK293 cells iron chelator deferoxamine (DFO, also known as desferal) decreased human α-synuclein mRNA levels [105]. Together with the evidence of an IRE presence in the 5′-UTR of the Alzheimer's amyloid precursor protein (APP) transcript [106], these evidence suggests that iron might contribute to neurodegenerative disease course by up-regulating α-synuclein or APP levels.

Alpha-synuclein undergoes a variety of post-translational modifications, which have been linked to the aggregation and pathology of α-synuclein [107, 108]. Nitration and phosphorylation were believed to be associated with iron and iron-induced oxidative stress. Nitrated α-synuclein is present in LBs, as well as in the insoluble fractions of affected brain regions in PD [109]. It is suggested that nitration of only a single tyrosine residue is sufficient to induce profound changes in catalytic activities and structural conformations of proteins. The attachment of a nitro molecule to α-synuclein at tyrosine residues (Y39, Y125, Y133, and Y136), is sufficient to induce profound changes in α-synuclein [110]. Nitrated of α-synuclein monomers and dimers could be mixed into the fibrils and accelerate the rate of fibril formation of unmodified α-synuclein [111, 112]. It was demonstrated that nitrated monomeric α-synuclein was degraded at a slower rate; together with diminished binding to lipid vesicles, nitrated α-synuclein were responsible for increasing fibrils formation and intracytoplasmic inclusions [113]. Excessive iron levels are associated with the increase of oxidative/nitrative stress, leading to elevated levels of tyrosine nitration. Iron supplements in rats markedly increased the extent of lipid peroxidation, protein carbonyls tyrosine nitration and oxidative metabolism of NO in liver [114]. There are currently no reports on the effects of iron on α-synuclein nitration. However, oxidative stress, the most common deleterious condition induced by iron, was believed to be involved in the nitrated process of α-synuclein. In vitro evidence demonstrated that microglial activation could induce NO-dependent oxidative stress in dopaminergic neurons and then cause α-synuclein nitration. These nitrated, aggregated α-synuclein during oxidative stress responses, in turn, incurs inflammatory microglial functions [115, 116]. In vivo, nitrated-α-synuclein (Tyr125, Tyr133) levels was elevated in the thymus of MPTP induced PD mice; nitrated recombinant α-synuclein was able to activate microglia and astrocytes lead to the degeneration of dopaminergic neurons [117, 118].

Phosphorylation of α-synuclein represents more than 90% of the total α-synuclein found in LB, whereas only 4% of the soluble monomeric α-synuclein is phosphorylated physiologically [119]. Phosphorylation at S129 is the major post-translational modification of α-synuclein [120]. Genetic approaches mimicking phosphorylated α-synuclein at S129 was associated with pathology in a transgenic Drosophila model, as well as in MPTP-treated monkeys. The mutation of serine 129 to aspartate significantly promotes α-synuclein phosphorylation and aggregation and thus dopaminergic neurons death. However, altering serine 129 to alanine prevents phosphorylation and suppresses dopaminergic neuronal loss [121]. Phosphorylated S129 α-synuclein aggregation was even correlated with the extent of dopaminergic neuron loss in the SN, further supporting that phosphorylation of α-synuclein might be harmful in PD [122]. There is also contradictory evidence that phosphorylation has no accelerating effects on α-synuclein toxicity or even neuroprotective. Phosphorylation mimicking S129 phosphorylation did not cause motor dysfunction and growth retardation; or even reduce α-synuclein pathology, suppress dopaminergic neurodegeneration and restore motor impairments by promoting α-synuclein clearance [123–125]. Together, there is still no consensus on whether α-synuclein phosphorylation is promoting aggregation or degradation, or neurotoxic or neuroprotective [126]. However, there is clear evidence that iron-derived oxidative stress promotes phosphorylation. Iron overload in 3D5 cells caused inclusion formation and phosphorylated α-synuclein aggregation [127]. Similarly, iron caused α-synuclein phosphorylation at S129 in SH-SY5Y cells, which was linked to ROS formation and mitochondrial dysfunction [128]. Furthermore, the binding affinity between α-synuclein and iron might be altered by phosphorylation at either S129 or Y125. It was suggested that phosphorylation at S129 or Y125 increases the binding affinity for Fe^{2+} but not Fe^{3+} at C-terminal region of α-synuclein [129].

It was hypothesized that blocking the metal modulated α-synuclein translation might provide a potential pathway to ameliorate the pathologies of both metal and α-synuclein [130]. As well, targeting iron and α-synuclein interactions might be a plausible therapeutic strategy in PD.

4.4 Glial Contributions in Iron Metabolism Disturbance

Neurodegenerative processes trigger universal and conserved glial reactions represented by astroglial dysfunction and microglial activation. As the most abundant glial cell types in the brain, astroglia were physiologically thought of as passive support cells mopping up transmitters and maintaining extracellular ion levels, and thus necessary to ensure optimal neuronal functioning [131]. Astroglia in the SNpc were relatively low distributed compared to other brain region. In autopsies of the SN from PD patient, reactive astrocytosis is absent although the activation of microglia was consistently observed [132]. Less astroglial neuroprotective factors are likely to contribute to increased vulnerability to any PD insults. However, reactive astroglia may cause or contribute to neurodegenerative processes. Stressed/dysfunctional astroglia might cease to support the dopaminergic neurons, which in turn contributes to the degeneration. This is referred as a dual role of astroglia since they are capable of releasing ROS and pro-inflammatory factors, further exacerbating neuroinflammation and brain damage [133, 134]. As for microglia, as the resident innate immune cells in the brain, they were proposed to constantly move and monitor the area in which they reside for immune insults and invading pathogens. These were regarded as beneficial effects due to the clearance of damaged cells and pathogens, as well as detrimental effects by the release of neurotoxic factors. Among many factors related to PD pathology, microglia-mediated neuroinflammation is one of the most striking hallmarks [135–137].

Both astroglia and microglia contribute to well-balanced iron metabolism in the brain. They expressed almost all the iron transporters and distinct profiles of iron metabolism proteins [44, 138–140]. Astroglia do not have very much ferritin and iron, indicating that these cells provide little iron storage [132, 141, 142]. However, they are more resistant than neurons and other glial cells to iron-induced toxicity, which might be due to the large antioxidants system aiding in the protection, e.g., metallothioneins [143, 144]. In primary cultures enriched in neurons, astroglia, or microglia, microglia were found to be the most efficient in accumulating iron, followed by astroglia, and then neurons. Iron retention in microglia was three times higher than that in neurons when they were exposed to iron, suggesting microglia have the capacity to accumulate and store large quantities of iron [145]. In autopsies of the SN from PD patient, ferritin immune reactivity was observed in activated microglia, indicating the abundant ferritin levels might reflect the large iron storage capacity in microglia [132]. The relationship between the accumulation of iron and ferritin in microglia during neuroinflammation was not fully elucidated [146, 147]. However, iron-containing microglia/macrophages are consistently present in the specific regions where the neurodegeneration occurred, providing a possible explanation of the increasing concentration of iron [148].

Iron elevation is evident in astroglia and microglia in the SN of PD [149], which might be capable of buffering iron accumulation. Meanwhile, iron accumulation in the specific region inevitably aggravates astroglia and microglia activation, which are in turn relevant to iron deposit in the diseased state of PD. That means iron affects these neuroglia responses in PD, at the same time, the latter affects neuronal iron metabolism. Brain-derived neurotrophic factor (BDNF), which is mainly derived from astroglia to protect neurons against oxidative stress and apoptosis, act on intracellular IRPs and thus post-transcriptionally target iron importer DMT1 to ameliorate iron accumulation in neurons [150]. These could be the beneficial effects exerted by astroglia, both on normal conditions and in the diseased state of PD. Also, these results provide novel perspectives to view the advantages of neurotrophic factors in PD. Meanwhile, neurotoxin 6-OHDA, inducing iron accumulation in neurons, might also be capable of promoting iron transport rate in primary cultured ventral mesencephalic astroglia, indicating a different response between neurons and astroglia in PD [151]. It was supposed when applied to in vivo circumstances, astroglia are capable of transporting excess iron outside thus preventing themselves, their neighboring cells and even the regions from iron overload. More recently, lipocalin-2 (LCN2), a member of the highly heterogeneous secretory protein family of lipocalins, was reported to increase mainly in the reactive astroglia in both the SN and striatum in PD. Astroglial LCN2 up-regulation induced the nigrostriatal dopaminergic neurotoxicity was aggravated by iron repletion [152], reinforcing the associations among astroglia, iron, and neurodegeneration.

Only activated microglia contained both Lf and its messenger, indicating microglia are the Lf producing cells [58, 153]. As we discussed above, by binding to LfR, iron transport crossing the plasma membrane is achieved through iron-saturated Lf internalization by a receptor-mediated pathway. LfR immunoreactivity was increased and more pronounced in the region where the loss of dopaminergic neurons is severe, suggesting that increased LfR on vulnerable neurons may increase intraneuronal iron levels. However, Lf, either iron-saturated or not, may be protective because it might function as a mitochondrial calcium modulator or a specialized iron scavenger [154–156]. The mechanisms that how microglia released Lf influence other cells like neurons, especially in terms of iron metabolism in PD need to be established further. However, it is clear now that the pro-inflammatory cytokines, e.g., IL-1β and TNF-α release were significantly increased when microglia were activated. More striking, the release was enhanced in activated microglia with iron overload [157]. On the other hand, the activated microglia perturb neuronal iron homeostasis via these releasing cytokines. IL-1β/6 and TNF-α were proved to up-regulate iron importer-DMT1, whereas down-regulated iron exporter-FPN1, thus causing iron accumulation in neurons. Both intracellular IRPs and hepcidin levels contributed to this modulation [157, 158], thus provide powerful evidence that microglia might deteriorate iron-mediated neuropathologies in PD. This is meaningful to understand the detrimental situation with the presence of iron deposit and microglia activation at the foci of lesion.

Based on the fact that astroglia and microglia contribute to neuronal iron metabolism, as well as that their functions might be altered in the diseased status, we suppose that there is a feedback loop between glial contributions and neuronal iron metabolism disturbance. Therefore, modulating the factors, e.g., neurotrophic factors or deleterious pro-inflammatory cytokines, might be valuable targets in drug discovery, especially in the aspects of ameliorating iron metabolism disturbance in PD.

4.5 Iron Chelation as a Therapy for PD

The role of iron in the pathology of PD has logically led to the theory that iron chelation may be an effective therapy for prevention and treatment of the disease [4, 159–163]. Mice with low iron diets for six weeks may provide protection against MPTP-inducing insults [164]. Similarly, rats received an iron-deficient diet attenuated kainate-induced neurotoxicity [165]. The clinically iron chelators, DFO, deferiprone, and deferasirox, have been shown to attenuate 6-OHDA, MPTP or iron-induced oxidative stress, mitochondrial dysfunction and prevent α-synuclein aggregation both in vivo and in vitro [166–170]. Recent phase II clinical trial data also demonstrated that deferiprone improved motor symptoms of PD patients [41]. 8-hydroxyquinoline (8-HQ), a highly lipophilic complex, has been utilized as the base for VK-28, M30, clioquinol, and Q1. These iron chelators readily penetrate cell membranes and the blood-brain barrier and have been shown to be effective in both cell culture and the PD animal models [40, 171–175].

Recently, several novel iron chelators have been explored and synthesized. Lipophilic analogs of desferrioxamine B (conjugating ancillary compounds onto the amine terminus) attenuated neuronal loss in SK-N-BE2-M17 cell lines and MPTP-treated mice [176, 177]. A novel hybrid iron chelator, (-)-12 (D-607), rescued PC12 cells from toxicity induced by iron [178]. Iron chelation therapeutic nanoparticles protected by a zwitterionic poly(2-methacryloyloxyethyl phosphorylcholine) (PMPC), which delay the saturation of iron chelators in blood circulation and prolong the in vivo lifetime, provided a PD phenotype reversion therapy and significantly improved the living quality of the PD mice [179]. 7,8-dihydroxy-4-((methylamino)methyl)-2*H*-chromen-2-one (DHC12) chelated mitochondrial and cytoplasmic labile iron, protected mitochondria against lipid peroxidation, and antagonized MPTP-induced neuronal death [180]. Polysubstituted piperazine derivatives (a selection of 1-hydroxypyridin-2-one (1,2-HOPO) metal chelators) and PBT434 (an orally bioavailable 8-hydroxyquinazolin-4(3*H*)-one) have also been designed as new iron chelators and proved to be neuroprotective in dopaminergic cell lines and multiple models of PD [181–183].

It has been found that the extracts from many natural foods have iron-chelating properties. For example, epigallocatechin gallate (EGCG), a major polyphenol in green tea, regulated the iron-export protein FPN1 and exerted a neurorescuing effect in MPTP-induced PD mice and α-synuclein transfected-PC12 cells [184, 185]. Indeed, epidemiological evidence showed that drinking two cups of tea a day could reduce the risk of PD [186]. Several natural flavonoids, such as ginkgetin (a natural bioflavonoid isolated from leaves of Ginkgo biloba L), curcumin (1,7-bis[4-hydroxy 3-methoxy phenyl]-1,6-heptadiene-3,5-dione, the polyphenolic flavonoid from *Curcuma longa*), myricetin and quercetin (bioflavonoids found in abundance in fruits, vegetables, and herbs) exerted neuroprotective effects against neuroinjury in PD model induced by MPTP or 6-OHDA via chelating iron [187–190]. Baicalin, a major active ingredient of the plant *Scutellaria baicalensis* Georgi, has been shown to reduce iron accumulation and have protective effects on dopaminergic neurons in the SN of rotenone-induced PD rats, as well as in C6 cells [191, 192]. Ginsenosides, the principal active components of ginseng, also reduced nigral iron levels in MPTP/MPP^+-treated models via regulating the expression of DMT1 and FPN1 [21, 193, 194]. A major advantage of these natural chelators is the mixture of neuroprotective properties they possess. In addition to iron chelation, they have been shown to exert a variety of beneficial effects, including antioxidant, anti-inflammatory, anti-apoptosis, and anticancer effects. Therefore, this allows for a more comprehensive usage in the prevention and treatment of PD.

It is important to realize that whether and how iron chelators could remove excess iron from the specified brain areas, without affecting systemic iron homeostasis. Indeed, iron is an essential element in various physiological processes, e.g., DNA synthesis, mitochondrial respiration, and oxygen transport. Future studies should seek to identify more cellular specificity therapies to prevent and attenuate brain iron metabolism dysfunction in PD.

References

1. Dauer W, Przedborski S (2003) Parkinson's disease: mechanisms and models. Neuron 39(6):889–909
2. Parkinson J (2002) An essay on the shaking palsy. 1817. J Neuropsychiatry Clin Neurosci 14(2):223–236; discussion 222
3. Kalia LV, Lang AE (2015) Parkinson's disease. Lancet 386(9996):896–912
4. Weinreb O et al (2013) Targeting dysregulation of brain iron homeostasis in Parkinson's disease by iron chelators. Free Radic Biol Med 62:52–64
5. Gallagher DA, Lees AJ, Schrag A (2010) What are the most important nonmotor symptoms in patients with Parkinson's disease and are we missing them? Mov Disord 25(15):2493–2500
6. Antony PM et al (2013) The hallmarks of Parkinson's disease. FEBS J 280(23):5981–5993
7. Jiang H et al (2017) Brain iron metabolism dysfunction in Parkinson's disease. Mol Neurobiol 54(4):3078–3101
8. Surmeier DJ, Obeso JA, Halliday GM (2017) Selective neuronal vulnerability in Parkinson disease. Nat Rev Neurosci 18(2):101–113
9. Sofic E et al (1988) Increased iron (III) and total iron content in post mortem substantia nigra of parkinsonian brain. J Neural Transm 74(3):199–205
10. Dexter DT et al (1987) Increased nigral iron content in postmortem parkinsonian brain. Lancet 2(8569):1219–1220
11. Dexter DT et al (1989) Increased nigral iron content and alterations in other metal ions occurring in brain in Parkinson's disease. J Neurochem 52(6):1830–1836
12. Hirsch EC et al (1991) Iron and aluminum increase in the substantia nigra of patients with Parkinson's disease: an X-ray microanalysis. J Neurochem 56(2):446–451
13. Sofic E et al (1991) Selective increase of iron in substantia nigra zona compacta of parkinsonian brains. J Neurochem 56(3):978–982
14. Yu X et al (2013) Decreased iron levels in the temporal cortex in postmortem human brains with Parkinson disease. Neurology 80(5):492–495
15. Jia F et al (2018) High dietary iron supplement induces the nigrostriatal dopaminergic neurons lesion in transgenic mice expressing mutant A53T human α-synuclein. Front Aging Neurosci 10:97
16. Mizuno Y et al (1988) Inhibition of mitochondrial respiration by 1-methyl-4-phenyl-1,2,3,6-tetrahydropyridine (MPTP) in mouse brain in vivo. Neurosci Lett 91(3):349–353
17. Mochizuki H, Yasuda T (2012) Iron accumulation in Parkinson's disease. J Neural Transm (Vienna) 119(12):1511–1514
18. Mochizuki H et al (1994) Iron accumulation in the substantia nigra of 1-methyl-4-phenyl-1,2,3,6-tetrahydropyridine (MPTP)-induced hemiparkinsonian monkeys. Neurosci Lett 168(1–2):251–253
19. He Y et al (2003) Dopaminergic cell death precedes iron elevation in MPTP-injected monkeys. Free Radic Biol Med 35(5):540–547
20. Temlett JA et al (1994) Increased iron in the substantia nigra compacta of the MPTP-lesioned hemiparkinsonian African green monkey: evidence from proton microprobe elemental microanalysis. J Neurochem 62(1):134–146
21. Wang J et al (2009) Rg1 reduces nigral iron levels of MPTP-treated C57BL6 mice by regulating certain iron transport proteins. Neurochem Int 54(1):43–48
22. Jiang H, Qian ZM, Xie JX (2003) Increased DMT1 expression and iron content in MPTP-treated C57BL/6 mice. Sheng Li Xue Bao 55(5):571–576
23. Youdim MB (2003) What have we learnt from CDNA microarray gene expression studies about the role of iron in MPTP induced neurodegeneration and Parkinson's disease? J Neural Transm Suppl 65:73–88
24. Wang J, Jiang H, Xie JX (2004) Time dependent effects of 6-OHDA lesions on iron level and neuronal loss in rat nigrostriatal system. Neurochem Res 29(12):2239–2243

25. He Y, Lee T, Leong SK (1999) Time course of dopaminergic cell death and changes in iron, ferritin and transferrin levels in the rat substantia nigra after 6-hydroxydopamine (6-OHDA) lesioning. Free Radic Res 31(2):103–112
26. Zhu Y et al (2017) Iron accumulation and microglia activation contribute to substantia nigra hyperechogenicity in the 6-OHDA-induced rat model of Parkinson's disease. Parkinsonism Relat Disord 36:76–82
27. Olmedo-Diaz S et al (2017) An altered blood-brain barrier contributes to brain iron accumulation and neuroinflammation in the 6-OHDA rat model of Parkinson's disease. Neuroscience 362:141–151
28. Zhang S et al (2009) Up-regulation of divalent metal transporter 1 is involved in 1-methyl-4-phenylpyridinium (MPP(+))-induced apoptosis in MES23.5 cells. Neurobiol Agin 30(9):1466–1476
29. Xu H et al (2010) Rg1 protects iron-induced neurotoxicity through antioxidant and iron regulatory proteins in 6-OHDA-treated MES23.5 cells. J Cell Biochem 111(6):1537–1545
30. Zhang J et al (2005) Intrapallidal lipopolysaccharide injection increases iron and ferritin levels in glia of the rat substantia nigra and induces locomotor deficits. Neuroscience 135(3):829–838
31. Zhang Z et al (2014) Pro-inflammatory cytokine-mediated ferroportin down-regulation contributes to the nigral iron accumulation in lipopolysaccharide-induced Parkinsonian models. Neuroscience 257:20–30
32. Ben-Shachar D, Youdim MB (1991) Intranigral iron injection induces behavioral and biochemical "parkinsonism" in rats. J Neurochem 57(6):2133–2135
33. Nandipati S, Litvan I (2016) Environmental exposures and Parkinson's disease. Int J Environ Res Public Health 13(9)
34. Peng J, Oo ML, Andersen JK (2010) Synergistic effects of environmental risk factors and gene mutations in Parkinson's disease accelerate age-related neurodegeneration. J Neurochem 115(6):1363–1373
35. Xu J et al (2002) Dopamine-dependent neurotoxicity of α-synuclein: a mechanism for selective neurodegeneration in Parkinson disease. Nat Med 8(6):600–606
36. Hare DJ, Double KL (2016) Iron and dopamine: a toxic couple. Brain 139:1026–1035
37. Dias V, Junn E, Mouradian MM (2013) The role of oxidative stress in Parkinson's disease. J Parkinsons Dis 3(4):461–491
38. Stockwell BR et al (2017) Ferroptosis: a regulated cell death nexus linking metabolism, redox biology, and disease. Cell 171(2):273–285
39. Dixon SJ et al (2012) Ferroptosis: an iron-dependent form of nonapoptotic cell death. Cell 149(5):1060–1072
40. Kaur D et al (2003) Genetic or pharmacological iron chelation prevents MPTP-induced neurotoxicity in vivo: a novel therapy for Parkinson's disease. Neuron 37(6):899–909
41. Devos D et al (2014) Targeting chelatable iron as a therapeutic modality in Parkinson's disease. Antioxid Redox Signal 21(2):195–210
42. Trinder D, Baker E (2003) Transferrin receptor 2: a new molecule in iron metabolism. Int J Biochem Cell Biol 35(3):292–296
43. Lee DW, Andersen JK (2010) Iron elevations in the aging Parkinsonian brain: a consequence of impaired iron homeostasis? J Neurochem 112(2):332–339
44. Ward RJ et al (2014) The role of iron in brain ageing and neurodegenerative disorders. Lancet Neurol 13(10):1045–1060
45. Kalivendi SV et al (2003) 1-Methyl-4-phenylpyridinium (MPP^+)-induced apoptosis and mitochondrial oxidant generation: role of transferrin-receptor-dependent iron and hydrogen peroxide. Biochem J 371(Pt 1):151–164
46. Mastroberardino PG et al (2009) A novel transferrin/TfR2-mediated mitochondrial iron transport system is disrupted in Parkinson's disease. Neurobiol Dis 34(3):417–431
47. Lee PL et al (1998) The human Nramp2 gene: characterization of the gene structure, alternative splicing, promoter region and polymorphisms. Blood Cells Mol Dis 24(2):199–215
48. Hubert N, Hentze MW (2002) Previously uncharacterized isoforms of divalent metal transporter (DMT)-1: implications for regulation and cellular function. Proc Natl Acad Sci U S A 99(19):12345–12350

49. Song N et al (2007) Divalent metal transporter 1 up-regulation is involved in the 6-hydroxydopamine-induced ferrous iron influx. J Neurosci Res 85(14):3118–3126
50. Andrews NC (1999) The iron transporter DMT1. Int J Biochem Cell Biol 31(10):991–994
51. Roth JA et al (2010) Parkin regulates metal transport via proteasomal degradation of the 1B isoforms of divalent metal transporter 1. J Neurochem 113(2):454–464
52. Foot NJ et al (2008) Regulation of the divalent metal ion transporter DMT1 and iron homeostasis by a ubiquitin-dependent mechanism involving Ndfips and WWP2. Blood 112(10):4268–4275
53. Jiang H et al (2010) Up-regulation of divalent metal transporter 1 in 6-hydroxydopamine intoxication is IRE/IRP dependent. Cell Res 20(3):345–356
54. Salazar J et al (2008) Divalent metal transporter 1 (DMT1) contributes to neurodegeneration in animal models of Parkinson's disease. Proc Natl Acad Sci U S A 105(47):18578–18583
55. Jia W et al (2015) Ndfip1 attenuated 6-OHDA-induced iron accumulation via regulating the degradation of DMT1. Neurobiol Aging 36(2):1183–1193
56. Howitt J et al (2014) Increased Ndfip1 in the substantia nigra of Parkinsonian brains is associated with elevated iron levels. PLoS ONE 9(1):e87119
57. Baker EN et al (1994) Three-dimensional structure of lactoferrin in various functional states. Adv Exp Med Biol 357:1–12
58. Fillebeen C et al (2001) Lactoferrin is synthesized by activated microglia in the human substantia nigra and its synthesis by the human microglial CHME cell line is upregulated by tumor necrosis factor alpha or 1-methyl-4-phenylpyridinium treatment. Brain Res Mol Brain Res 96(1–2):103–113
59. Faucheux BA et al (1995) Expression of lactoferrin receptors is increased in the mesencephalon of patients with Parkinson disease. Proc Natl Acad Sci U S A 92(21):9603–9607
60. Wang J, Jiang H, Xie JX (2007) Ferroportin1 and hephaestin are involved in the nigral iron accumulation of 6-OHDA-lesioned rats. Eur J Neurosci 25(9):2766–2772
61. Yoshida K et al (1995) A mutation in the ceruloplasmin gene is associated with systemic hemosiderosis in humans. Nat Genet 9(3):267–272
62. Song N et al (2010) Ferroportin 1 but not hephaestin contributes to iron accumulation in a cell model of Parkinson's disease. Free Radic Biol Med 48(2):332–341
63. Boll MC et al (1999) Reduced ferroxidase activity in the cerebrospinal fluid from patients with Parkinson's disease. Neurosci Lett 265(3):155–158
64. Wang J, Bi M, Xie J (2015) Ceruloplasmin is involved in the nigral iron accumulation of 6-OHDA-lesioned rats. Cell Mol Neurobiol 35(5):661–668
65. Dragicevic E, Schiemann J, Liss B (2015) Dopamine midbrain neurons in health and Parkinson's disease: emerging roles of voltage-gated calcium channels and ATP-sensitive potassium channels. Neuroscience 284:798–814
66. Basso M et al (2004) Proteome analysis of human substantia nigra in Parkinson's disease. Proteomics 4(12):3943–3952
67. Tsushima RG et al (1999) Modulation of iron uptake in heart by L-type Ca^{2+} channel modifiers: possible implications in iron overload. Circ Res 84(11):1302–1309
68. Oudit GY et al (2003) L-type Ca^{2+} channels provide a major pathway for iron entry into cardiomyocytes in iron-overload cardiomyopathy. Nat Med 9(9):1187–1194
69. Gaasch JA et al (2007) Voltage-gated calcium channels provide an alternate route for iron uptake in neuronal cell cultures. Neurochem Res 32(10):1686–1693
70. Ma Z, Zhou Y, Xie J (2012) Nifedipine prevents iron accumulation and reverses iron-overload-induced dopamine neuron degeneration in the substantia nigra of rats. Neurotox Res 22(4):274–279
71. Carvajal FJ, Mattison HA, Cerpa W (2016) Role of NMDA receptor-mediated glutamatergic signaling in chronic and acute neuropathologies. Neural Plast 2016:2701526
72. Ambrosi G, Cerri S, Blandini F (2014) A further update on the role of excitotoxicity in the pathogenesis of Parkinson's disease. J Neural Transm (Vienna) 121(8):849–859
73. Schiemann J et al (2012) K-ATP channels in dopamine substantia nigra neurons control bursting and novelty-induced exploration. Nat Neurosci 15(9):1272–1280

74. Nakamichi N et al (2004) Relevant modulation by ferrous ions of N-methyl-D-aspartate receptors in ischemic brain injuries. Curr Neurovasc Res 1(5):429–440
75. Pelizzoni I et al (2011) Iron handling in hippocampal neurons: activity-dependent iron entry and mitochondria-mediated neurotoxicity. Aging Cell 10(1):172–183
76. Cheah JH et al (2006) NMDA receptor-nitric oxide transmission mediates neuronal iron homeostasis via the GTPase Dexras1. Neuron 51(4):431–440
77. Chen Y et al (2013) Dexras1, a small GTPase, is required for glutamate-NMDA neurotoxicity. J Neurosci 33(8):3582–3587
78. Guix FX et al (2005) The physiology and pathophysiology of nitric oxide in the brain. Prog Neurobiol 76(2):126–152
79. Chen Y et al (2015) PKA-mediated phosphorylation of Dexras1 suppresses iron trafficking by inhibiting S-nitrosylation. FEBS Lett 589(20 Pt B):3212–3219
80. Choi BR et al (2013) PKA modulates iron trafficking in the striatum via small GTPase. Rhes. Neuroscience 253:214–220
81. White RS et al (2016) Lysosomal iron modulates NMDA receptor-mediated excitation via small GTPase, Dexras1. Mol Brain 9:38
82. Xu H et al (2018) Activation of NMDA receptors mediated iron accumulation via modulating iron transporters in Parkinson's disease. FASEB J, fj201800060RR
83. Haeger P et al (2010) Increased hippocampal expression of the divalent metal transporter 1 (DMT1) mRNA variants 1B and +IRE and DMT1 protein after NMDA-receptor stimulation or spatial memory training. Neurotox Res 17(3):238–247
84. Liss B, Roeper J (2001) ATP-sensitive potassium channels in dopaminergic neurons: transducers of mitochondrial dysfunction. News Physiol Sci 16:214–217
85. Liss B et al (2005) K-ATP channels promote the differential degeneration of dopaminergic midbrain neurons. Nat Neurosci 8(12):1742–1751
86. Li YX, Bertram R, Rinzel J (1996) Modeling N-methyl-D-aspartate-induced bursting in dopamine neurons. Neuroscience 71(2):397–410
87. Chan CS et al (2007) 'Rejuvenation' protects neurons in mouse models of Parkinson's disease. Nature 447(7148):1081–1086
88. Guzman JN et al (2010) Oxidant stress evoked by pacemaking in dopaminergic neurons is attenuated by DJ-1. Nature 468(7324):696–700
89. Gunshin H et al (1997) Cloning and characterization of a mammalian proton-coupled metal-ion transporter. Nature 388(6641):482–488
90. Du X et al (2016) Activation of ATP-sensitive potassium channels enhances DMT1-mediated iron uptake in SK-N-SH cells in vitro. Sci Rep 6:33674
91. Takanashi M et al (2001) Iron accumulation in the substantia nigra of autosomal recessive juvenile parkinsonism (ARJP). Parkinsonism Relat Disord 7(4):311–314
92. Altamura S, Muckenthaler MU (2009) Iron toxicity in diseases of aging: Alzheimer's disease, Parkinson's disease and atherosclerosis. J Alzheimer's Dis 16(4):879–895
93. Curtis AR et al (2001) Mutation in the gene encoding ferritin light polypeptide causes dominant adult-onset basal ganglia disease. Nat Genet 28(4):350–354
94. Harris ZL et al (1995) Aceruloplasminemia: molecular characterization of this disorder of iron metabolism. Proc Natl Acad Sci U S A 92(7):2539–2543
95. Nichols CG (2006) KATP channels as molecular sensors of cellular metabolism. Nature 440(7083):470–476
96. Holemans S et al (1994) Sulfonylurea binding sites in normal human brain and in Parkinson's disease, progressive supranuclear palsy and Huntington's disease. Brain Res 642(1–2):327–333
97. Zhang CW et al (2017) Transgenic mice overexpressing the divalent metal transporter 1 exhibit iron accumulation and enhanced Parkin expression in the brain. Neuromolecular Med 19(2–3):375–386
98. Castellani RJ et al (2000) Sequestration of iron by Lewy bodies in Parkinson's disease. Acta Neuropathol 100(2):111–114

99. Binolfi A et al (2006) Interaction of α-synuclein with divalent metal ions reveals key differences: a link between structure, binding specificity and fibrillation enhancement. J Am Chem Soc 128(30):9893–9901
100. Uversky VN, Li J, Fink AL (2001) Metal-triggered structural transformations, aggregation, and fibrillation of human α-synuclein. A possible molecular NK between Parkinson's disease and heavy metal exposure. J Biol Chem 276(47):44284–4496
101. Bharathi, Indi SS, Rao KS (2007) Copper- and iron-induced differential fibril formation in α-synuclein: TEM study. Neurosci Lett 424(2):78–82
102. Friedlich AL, Tanzi RE, Rogers JT (2007) The 5'-untranslated region of Parkinson's disease α-synuclein messengerRNA contains a predicted iron responsive element. Mol Psychiatry 12(3):222–223
103. Li W et al (2011) Oxidative stress partially contributes to iron-induced α-synuclein aggregation in SK-N-SH cells. Neurotox Res 19(3):435–442
104. Koukouraki P, Doxakis E (2016) Constitutive translation of human α-synuclein is mediated by the 5'-untranslated region. Open Biol 6(4):160022
105. Febbraro F et al (2012) α-Synuclein expression is modulated at the translational level by iron. NeuroReport 23(9):576–580
106. Rogers JT et al (2002) An iron-responsive element type II in the 5'-untranslated region of the Alzheimer's amyloid precursor protein transcript. J Biol Chem 277(47):45518–45528
107. Oueslati A, Fournier M, Lashuel HA (2010) Role of post-translational modifications in modulating the structure, function and toxicity of α-synuclein: implications for Parkinson's disease pathogenesis and therapies. Prog Brain Res 183:115–145
108. Barrett PJ,. Timothy Greenamyre, J (2015) Post-translational modification of α-synuclein in Parkinson's disease. Brain Res 1628(Pt B):247–253
109. Giasson BI et al (2000) Oxidative damage linked to neurodegeneration by selective α-synuclein nitration in synucleinopathy lesions. Science 290(5493):985–989
110. Reynolds MR, Berry RW, Binder LI (2007) Nitration in neurodegeneration: deciphering the "Hows" "nYs". Biochemistry 46(25):7325–7336
111. Burai R et al (2015) Elucidating the role of site-SPECIFIC Nitration of α-Synuclein in the pathogenesis of Parkinson's disease via protein semisynthesis and mutagenesis. J Am Chem Soc 137(15):5041–5052
112. Kleinknecht A et al (2016) C-terminal tyrosine residue modifications modulate the protective phosphorylation of serine 129 of α-synuclein in a yeast model of Parkinson's disease. PLoS Genet 12(6):e1006098
113. Hodara R et al (2004) Functional consequences of α-synuclein tyrosine nitration: diminished binding to lipid vesicles and increased fibril formation. J Biol Chem 279(46):47746–47753
114. Li X et al (2012) Iron increases liver injury through oxidative/nitrative stress in diabetic rats: involvement of nitrotyrosination of glucokinase. Biochimie 94(12):2620–2627
115. Shavali S, Combs CK, Ebadi M (2006) Reactive macrophages increase oxidative stress and α-synuclein nitration during death of dopaminergic neuronal cells in co-culture: relevance to Parkinson's disease. Neurochem Res 31(1):85–94
116. Reynolds AD et al (2008) Nitrated α-synuclein-activated microglial profiling for Parkinson's disease. J Neurochem 104(6):1504–1525
117. Wen L et al (2018) NLRP3 inflammasome activation in the thymus of MPTP-induced Parkinsonian mouse model. Toxicol Lett 288:1–8
118. Yu Z et al (2010) Nitrated α-synuclein induces the loss of dopaminergic neurons in the substantia nigra of rats. PLoS ONE 5(4):e9956
119. Fujiwara H et al (2002) α-Synuclein is phosphorylated in synucleinopathy lesions. Nat Cell Biol 4(2):160–164
120. Anderson JP et al (2006) Phosphorylation of Ser-129 is the dominant pathological modification of α-synuclein in familial and sporadic Lewy body disease. J Biol Chem 281(40):29739–29752
121. Chen L, Feany MB (2005) α-synuclein phosphorylation controls neurotoxicity and inclusion formation in a Drosophila model of Parkinson disease. Nat Neurosci 8(5):657–663

122. Huang B et al (2018) Phosphorylated α-synuclein accumulations and Lewy body-like pathology distributed in Parkinson's disease-related brain areas of aged Rhesus monkeys treated with MPTP. Neuroscience 379:302–315
123. Kuwahara T et al (2012) Phosphorylation of α-synuclein protein at Ser-129 reduces neuronal dysfunction by lowering its membrane binding property in Caenorhabditis elegans. J Biol Chem 287(10):7098–7109
124. Oueslati A et al (2013) Polo-like kinase 2 regulates selective autophagic α-synuclein clearance and suppresses its toxicity in vivo. Proc Natl Acad Sci U S A 110(41):E3945–E3954
125. Sancenon V et al (2012) Suppression of α-synuclein toxicity and vesicle trafficking defects by phosphorylation at S129 in yeast depends on genetic context. Hum Mol Genet 21(11):2432–2449
126. Tenreiro S, Eckermann K, Outeiro TF (2014) Protein phosphorylation in neurodegeneration: friend or foe? Front Mol Neurosci 7:42
127. Takahashi M et al (2007) Oxidative stress-induced phosphorylation, degradation and aggregation of α-synuclein are linked to upregulated CK2 and cathepsin D. Eur J Neurosci 26(4):863–874
128. Perfeito R et al (2014) Linking α-synuclein phosphorylation to reactive oxygen species formation and mitochondrial dysfunction in SH-SY5Y cells. Mol Cell Neurosci 62:51–59
129. Lu Y et al (2011) Phosphorylation of α-Synuclein at Y125 and S129 alters its metal binding properties: implications for understanding the role of α-Synuclein in the pathogenesis of Parkinson's Disease and related disorders. ACS Chem Neurosci 2(11):667–675
130. Cahill CM et al (2009) Amyloid precursor protein and alpha synuclein translation, implications for iron and inflammation in neurodegenerative diseases. Biochim Biophys Acta 1790(7):615–628
131. Sofroniew MV, Vinters HV (2010) Astrocytes: biology and pathology. Acta Neuropathol 119(1):7–35
132. Mirza B et al (2000) The absence of reactive astrocytosis is indicative of a unique inflammatory process in Parkinson's disease. Neuroscience 95(2):425–432
133. Heneka MT, Rodriguez JJ, Verkhratsky A (2010) Neuroglia in neurodegeneration. Brain Res Rev 63(1–2):189–211
134. Neal M, Richardson JR (2017) Epigenetic regulation of astrocyte function in neuroinflammation and neurodegeneration. Biochim Biophys Acta 1864(2):432–443
135. Collins LM et al (2012) Contributions of central and systemic inflammation to the pathophysiology of Parkinson's disease. Neuropharmacology 62(7):2154–2168
136. Long-Smith CM, Sullivan AM, Nolan YM (2009) The influence of microglia on the pathogenesis of Parkinson's disease. Prog Neurobiol 89(3):277–287
137. Phani S, Loike JD, Przedborski S (2012) Neurodegeneration and inflammation in Parkinson's disease. Parkinsonism Relat Disord 18(Suppl 1):S207–S209
138. Benarroch EE (2009) Brain iron homeostasis and neurodegenerative disease. Neurology 72(16):1436–1440
139. Rouault TA, Zhang DL, Jeong SY (2009) Brain iron homeostasis, the choroid plexus, and localization of iron transport proteins. Metab Brain Dis 24(4):673–684
140. Healy S, McMahon JM, FitzGerald U (2017) Modelling iron mismanagement in neurodegenerative disease in vitro: paradigms, pitfalls, possibilities & practical considerations. Prog Neurobiol 158:1–14
141. Zecca L et al (2004) Iron, brain ageing and neurodegenerative disorders. Nat Rev Neurosci 5(11):863–873
142. Bartzokis G et al (2007) Brain ferritin iron may influence age- and gender-related risks of neurodegeneration. Neurobiol Aging 28(3):414–423
143. Kress GJ, Dineley KE, Reynolds IJ (2002) The relationship between intracellular free iron and cell injury in cultured neurons, astrocytes, and oligodendrocytes. J Neurosci 22(14):5848–5855
144. Oshiro S et al (2008) Microglia and astroglia prevent oxidative stress-induced neuronal cell death: implications for aceruloplasminemia. Biochim Biophys Acta 1782(2):109–117

145. Bishop GM et al (2011) Accumulation of non-transferrin-bound iron by neurons, astrocytes, and microglia. Neurotox Res 19(3):443–451
146. Rathnasamy G, Ling EA, Kaur C (2013) Consequences of iron accumulation in microglia and its implications in neuropathological conditions. CNS Neurol Disord: Drug Targets 12(6):785–798
147. Thomsen MS et al (2015) Neurodegeneration with inflammation is accompanied by accumulation of iron and ferritin in microglia and neurons. Neurobiol Dis 81:108–118
148. Andersen HH, Johnsen KB, Moos T (2014) Iron deposits in the chronically inflamed central nervous system and contributes to neurodegeneration. Cell Mol Life Sci 71(9):1607–1622
149. Jellinger K et al (1990) Brain iron and ferritin in Parkinson's and Alzheimer's diseases. J Neural Transm Park Dis Dement Sect 2(4):327–340
150. Zhang HY et al (2014) Brain-derived neurotrophic factor and glial cell line-derived neurotrophic factor inhibit ferrous iron influx via divalent metal transporter 1 and iron regulatory protein 1 regulation in ventral mesencephalic neurons. Biochim Biophys Acta 1843(12):2967–2975
151. Zhang HY et al (2013) 6-hydroxydopamine promotes iron traffic in primary cultured astrocytes. Biometals 26(5):705–714
152. Kim BW et al (2016) Pathogenic upregulation of glial lipocalin-2 in the Parkinsonian dopaminergic system. J Neurosci 36(20):5608–5622
153. An L et al (2009) Expression and localization of lactotransferrin messenger RNA in the cortex of Alzheimer's disease. Neurosci Lett 452(3):277–280
154. Fillebeen C et al (1999) Lactoferrin is synthesized by mouse brain tissue and its expression is enhanced after MPTP treatment. Brain Res Mol Brain Res 72(2):183–194
155. Rousseau E, Michel PP, Hirsch EC (2013) The iron-binding protein lactoferrin protects vulnerable dopamine neurons from degeneration by preserving mitochondrial calcium homeostasis. Mol Pharmacol 84(6):888–898
156. Wang J et al (2015) The protective effect of lactoferrin on ventral mesencephalon neurons against MPP^+ is not connected with its iron binding ability. Sci Rep 5:10729
157. Wang J et al (2013) Pro-inflammatory cytokines modulate iron regulatory protein 1 expression and iron transportation through reactive oxygen/nitrogen species production in ventral mesencephalic neurons. Biochim Biophys Acta 1832(5):618–625
158. Urrutia P et al (2013) Inflammation alters the expression of DMT1, FPN1 and hepcidin, and it causes iron accumulation in central nervous system cells. J Neurochem 126(4):541–549
159. Kolnagou A, Kontoghiorghe CN, Kontoghiorghes GJ (2018) New targeted therapies and diagnostic methods for iron overload diseases. Front Biosci (Schol Ed) 10:1–20
160. Dusek P, Schneider SA, Aaseth J (2016) Iron chelation in the treatment of neurodegenerative diseases. J Trace Elem Med Biol 38:81–92
161. Gumienna-Kontecka E et al (2014) Iron chelating strategies in systemic metal overload, neurodegeneration and cancer. Curr Med Chem 21(33):3741–3767
162. Mounsey RB, Teismann P (2012) Chelators in the treatment of iron accumulation in Parkinson's disease. Int J Cell Biol 2012:983245
163. Hider RC et al (2011) The potential application of iron chelators for the treatment of neurodegenerative diseases. Metallomics 3(3):239–249
164. Levenson CW et al (2004) Role of dietary iron restriction in a mouse model of Parkinson's disease. Exp Neurol 190(2):506–514
165. Shoham S, Youdim MB (2004) Nutritional iron deprivation attenuates kainate-induced neurotoxicity in rats: implications for involvement of iron in neurodegeneration. Ann N Y Acad Sci 1012:94–114
166. Sangchot P et al (2002) Deferoxamine attenuates iron-induced oxidative stress and prevents mitochondrial aggregation and α-synuclein translocation in SK-N-SH cells in culture. Dev Neurosci 24(2–3):143–153
167. Jiang H et al (2006) Neuroprotective effects of iron chelator Desferal on dopaminergic neurons in the substantia nigra of rats with iron-overload. Neurochem Int 49(6):605–609

168. Lan J, Jiang DH (1997) Desferrioxamine and vitamin E protect against iron and MPTP-induced neurodegeneration in mice. J Neural Transm (Vienna) 104(4–5):469–481
169. Febbraro F et al (2013) Chronic intranasal deferoxamine ameliorates motor defects and pathology in the α-synuclein rAAV Parkinson's model. Exp Neurol 247:45–58
170. Dexter DT et al (2011) Clinically available iron chelators induce neuroprotection in the 6-OHDA model of Parkinson's disease after peripheral administration. J Neural Transm (Vienna) 118(2):223–231
171. Shachar DB et al (2004) Neuroprotection by a novel brain permeable iron chelator, VK-28, against 6-hydroxydopamine lession in rats. Neuropharmacology 46(2):254–263
172. Gal S et al (2010) Restoration of nigrostriatal dopamine neurons in post-MPTP treatment by the novel multifunctional brain-permeable iron chelator-monoamine oxidase inhibitor drug, M30. Neurotox Res 17(1):15–27
173. Mena NP et al (2015) The novel mitochondrial iron chelator 5-((methylamino)methyl)-8-hydroxyquinoline protects against mitochondrial-induced oxidative damage and neuronal death. Biochem Biophys Res Commun 463(4):787–792
174. Zhu W et al (2007) Prevention and restoration of lactacystin-induced nigrostriatal dopamine neuron degeneration by novel brain-permeable iron chelators. FASEB J 21(14):3835–3844
175. Gal S et al (2006) M30, a novel multifunctional neuroprotective drug with potent iron chelating and brain selective monoamine oxidase-ab inhibitory activity for Parkinson's disease. J Neural Transm Suppl 70:447–456
176. Gotsbacher MP et al (2017) Analogues of desferrioxamine B designed to attenuate iron-mediated neurodegeneration: synthesis, characterisation and activity in the MPTP-mouse model of Parkinson's disease. Metallomics 9(7):852–864
177. Telfer TJ et al (2017) Adamantyl- and other polycyclic cage-based conjugates of desferrioxamine B (DFOB) for treating iron-mediated toxicity in cell models of Parkinson's disease. Bioorg Med Chem Lett 27(8):1698–1704
178. Das B et al (2017) A Novel Iron(II) Preferring Dopamine Agonist Chelator as Potential Symptomatic and Neuroprotective Therapeutic Agent for Parkinson's Disease. ACS Chem Neurosci 8(4):723–730
179. Wang N et al (2017) Iron chelation nanoparticles with delayed saturation as an effective therapy for Parkinson disease. Biomacromol 18(2):461–474
180. Aguirre P et al (2017) Neuroprotective Effect of a new 7,8-dihydroxycoumarin-based Fe(2+)/Cu(2+) chelator in cell and animal models of Parkinson's disease. ACS Chem Neurosci 8(1):178–185
181. Abdelsayed S et al (2015) Piperazine derivatives as iron chelators: a potential application in neurobiology. Biometals 28(6):1043–1061
182. Workman DG et al (2015) Protection from neurodegeneration in the 6-hydroxydopamine (6-OHDA) model of Parkinson's with novel 1-hydroxypyridin-2-one metal chelators. Metallomics 7(5):867–876
183. Finkelstein DI et al (2017) The novel compound PBT434 prevents iron mediated neurodegeneration and α-synuclein toxicity in multiple models of Parkinson's disease. Acta Neuropathol Commun 5(1):53
184. Xu Q et al (2017) Epigallocatechin gallate has a neurorescue effect in a mouse model of Parkinson disease. J Nutr 147(10):1926–1931
185. Zhao J et al (2017) Metal chelator EGCG attenuates Fe(III)-induced conformational transition of α-synuclein and protects AS-PC12 cells against Fe(III)-induced death. J Neurochem 143(1):136–146
186. Chan DK et al (1998) Genetic and environmental risk factors for Parkinson's disease in a Chinese population. J Neurol Neurosurg Psychiatry 65(5):781–784
187. Wang YQ et al (2015) Neuroprotective effects of ginkgetin against neuroinjury in Parkinson's disease model induced by MPTP via chelating iron. Free Radic Res 49(9):1069–1080
188. Ma ZG et al (2007) Myricetin reduces 6-hydroxydopamine-induced dopamine neuron degeneration in rats. NeuroReport 18(11):1181–1185

189. Haleagrahara N, Siew CJ, Ponnusamy K (2013) Effect of quercetin and desferrioxamine on 6-hydroxydopamine (6-OHDA) induced neurotoxicity in striatum of rats. J Toxicol Sci 38(1):25–33
190. Du XX et al (2012) Curcumin protects nigral dopaminergic neurons by iron-chelation in the 6-hydroxydopamine rat model of Parkinson's disease. Neurosci Bull 28(3):253–258
191. Guo C, Chen X, Xiong P (2014) Baicalin suppresses iron accumulation after substantia nigra injury: relationship between iron concentration and transferrin expression. Neural Regen Res 9(6):630–636
192. Xiong P et al (2012) Baicalin and deferoxamine alleviate iron accumulation in different brain regions of Parkinson's disease rats. Neural Regen Res 7(27):2092–2098
193. Xu H et al (2010) Rg1 protects the MPP^+ -treated MES23.5 cells via attenuating DMT1 up-regulation and cellular iron uptake. Neuropharmacology 58(2):488–494
194. Gonzalez-Burgos E, Fernandez-Moriano C, Gomez-Serranillos MP (2015) Potential neuroprotective activity of Ginseng in Parkinson's disease: a review. J Neuroimmune Pharmacol 10(1):14–29

Chapter 5 Iron Pathophysiology in Alzheimer's Diseases

Tao Wang, Shuang-Feng Xu, Yong-Gang Fan, Lin-Bo Li and Chuang Guo

Abstract Alzheimer's disease (AD) is a multifactorial neurodegenerative condition associated with pathological accumulation of amyloid plaques and with the appearance of deposit of neurofibrillary tangles. Increasing evidence suggests that disorders of metal ion metabolism in the brain are one of the risk factors for the pathogenesis of AD. Iron, one of the endogenous metal ions, involves in many important physiological activities in the brain. Iron metabolism mainly depends on iron regulatory proteins including ferritin, transferrin and transferrin receptor, hepcidin, ferroportin, lactoferrin. Abnormal iron metabolism generates hydroxyl radicals through the Fenton reaction, triggers oxidative stress reactions, damages cell lipids, protein and DNA structure and function, leads to cell death, and ultimately influences the process of β-amyloid (Aβ) misfolding and plaque aggregation. Although the results are different, in general, iron has deposition in different brain regions of AD patients, which may impair normal cognitive function and behavior. Therefore, neuroimaging changes have so far been largely attributed to focal iron deposition accompanying the plaques at preclinical stages of AD, and iron-targeted therapeutic strategies have become a new direction. Iron chelators have received a great deal of attention and have obtained good results in scientific experiments and some clinical trials. Future research will also focus on iron as an opportunity to study the mechanism of the occurrence and development of AD from the iron steady state to more fully clarify the etiology and prevention strategies.

Keywords Alzheimer's disease · Iron · Oxidative stress · Mitochondria · Chelators

---

Tao Wang, Shuang-Feng Xu, Yong-Gang Fan, Lin-Bo Li—These authors contributed equally to this work

---

T. Wang · S.-F. Xu · Y.-G. Fan · L.-B. Li · C. Guo (✉)
College of Life and Health Sciences, Northeastern University, No. 195, Chuangxin Road, Hunnan District, Shenyang 110169, China
e-mail: guoc@mail.neu.edu.cn

Y.-Z. Chang (ed.), *Brain Iron Metabolism and CNS Diseases*,
Advances in Experimental Medicine and Biology 1173,
https://doi.org/10.1007/978-981-13-9589-5_5

5.1 Introduction

Alzheimer's disease (AD) is a chronic neurodegenerative disease that occurs mostly in older people over the age of 65 [167]. In 1906, the German psychotic and neuropathologist Alois Alzheimer first described the disease that was later to carry his name as a syndrome that involves progressive cognitive decline and behavioral changes underpinned by senile plaques (SPs) and neurofibrillary tangles (NFTs) in the autopsied brains of people with AD [90]. As the society ages, AD is the main cause of dementia and one of the great healthcare challenges of the twenty-first century [246]. According to the World Alzheimer's Disease Report of 2018 issued by the International Alzheimer's Association (ADI), there are about 50 million dementia patients in the world in 2018, mainly dementia caused by AD. According to the prevalence of AD and the aging process of society, by 2050, the number of dementia patients worldwide is estimated to reach 152 million, equivalent to the size of Russia or Bangladesh. At the same time, the current global cost of preventing and treating dementia is about $1 trillion, and it is expected to double by 2030 [297]. However, to date, no cure or therapeutic intervention is available for AD that can treat the underlying cause of AD or slow disease progression [139]. It can be seen that AD not only seriously affects the quality of life of the elderly, but also increases the burden on families and society, making it a huge challenge for the medical community and society [9, 167].

5.2 Clinical and Pathological Hallmarks of Alzheimer Disease

Clinically, AD presents as familial AD (fAD) or sporadic AD (sAD). Although the vast majority of AD is sporadic, the presence of mutations in one of three genes—amyloid precursor protein (APP), presenilin 1 (PSEN1), and presenilin 2 (PSEN2)—can lead to rare (<0.5%) familial form of AD [22]. Symptoms of fAD develop earlier than sporadic AD, usually between 30 and 50 years of age [163]. The clinical symptoms of AD are mainly cognitive decline, decreased ability to live and abnormal mental behavior. According to recommendations from the working groups convened by National Institute of Aging and the Alzheimer's Association in 2011, the dementia stages of AD were divided into the preclinical/presymptomatic stage of AD, the symptomatic predementia stage called mild cognitive impairment (MCI) due to AD, and AD dementia [9, 261]. At present, the MCI can be found through neuropsychological testing, which can be used as a preliminary clinical diagnostic criterion for AD [38, 52]. The neuropsychological characteristics of MCI are mainly memory impairment. In fact, memory impairment is an important change in brain function during normal aging, and the core symptoms and pre-early characteristic symptoms of AD are also memory disorders [261]. Thus, it can be said that memory impairment (memory loss) runs through normal aging to pre-early AD. Similar to the pre-symptoms, early symptoms of AD are also memory disorders, especially near-memory disorders; that is, short-term events cannot be recalled. As the pathology of AD progresses, the patient's personality and behavior will change, severe cognitive function declines, mental disorders appear, speaking and acting very difficult, and losing self-care ability in the late stage, and following extremely high mortality from multiple complications [9].

Pathologically, the typical pathological features of AD, the deposition of NFTs formed by hyperphosphorylation of tau protein in neurons and extracellular amyloid-β (Aβ) plaques, can be detected in early AD [167]. NFTs and Aβ are also detected on PET scanning in healthy human brains, but normal organisms metabolize them or allow them to dissolve in cell fluids to perform normal functions [167]. In the brain of AD patients, these proteins are excessively present, and the clearance is blocked, which affects cell survival along with synaptic and neuronal losses particularly in cortex and hippocampus [9, 90] and hence atrophy of brain tissue. Hippocampus is associated with memory function; hippocampal atrophy is regarded as an important indicator of memory impairment. When performing neurological examination and cranial CT or magnetic resonance imaging (MRI) in early AD, most of them have positive findings [167]. As the pathology of AD progresses, the plaque and tangles formed by the middle of the brain in the memory and thinking area of the AD patients are more, resulting in impaired memory and thinking ability of the patient. At this stage, plaque and tangles spread to the sensory area responsible for speaking and understanding of the language and for the body to be in contact with the surrounding objects. Its pathological manifestations are atrophy of the cerebral cortex, flattening of the brain, widening of the sulcus, enlargement of the ventricles, and weight reduction of brain. The atrophy of the temporal lobe, parietal lobe, prefrontal cortex, and hippocampus was the most obvious, and it was more significant than the early patients. In the advanced stage of AD, most of the patient's cerebral cortex has been severely damaged, neurons have died, and the brain has severely shrunk [221, 282].

5.3 Iron Dyshomeostasis and AD

The etiology of AD is complex and diverse, and it is the result of a combination of genetics, environment, and life, but the specific mechanism of its onset is still unclear, and there is no corresponding treatment. Several hypotheses of AD pathology have been discovered including extracellular deposition of Aβ, intracellular accumulation of hyperphosphorylated tau, mitochondrial structure and function disorders, oxidative stress, disorders of metal ion metabolism, damage to N-methyl-D-aspartate receptor (NMDAR)-associated signaling pathways, abnormal lipid metabolism, and abnormal cell cycle in some neurons [39, 226]. Note that metals are essential for life and play a central role in many biochemical pathways. Environmental exposure, aging, genetic dysfunction, inadequate dietary intake, and drug interaction can all induce an alteration in their homeostasis leading to deleterious effects and neurotoxicity [307]. In recent years, there is growing evidence that impaired iron homeostasis is an early event in the progression of AD [150, 176]. Abnormal iron content results in the loss of function of several enzymes that require iron as a cofactor, forming toxic oxidizing species, and increasing β-amyloid production [226]. Several common genetic polymorphisms that cause elevated levels of iron and disorders in the body are associated with AD [63, 198], but the pathology is unclear.

It is widely believed that aging is a major risk factor for neurodegenerative diseases. Iron will accumulate in several brain regions of healthy elderly people and will exhibit a large degree of accumulation in the brains of patients with neurodegenerative diseases [53, 55, 56, 119, 295]. In vitro, increasing iron concentration accelerates Aβ plaque and pTau tangles and increases its toxicity [240]. Autopsy results showed that total iron deposition was positively correlated with age, and altered iron levels were observed in specific brain areas of AD patients (the hippocampus) compared with the age-matched control group [232, 308]. Iron elevation in the AD brain was first reported in 1953 [108], and its association with Aβ and NFTs or with ferritin in peripheral glial cells was confirmed in some studies [55, 56, 150, 176, 295]. Clinical studies have also shown that the disorder of iron metabolism is the key to the pathology of AD. It is based on the observation that the patient's content in the cortex, subcortex, and white matter is significantly increased [233]. MRI analysis showed that iron levels in the hippocampus increased in the early stage of AD and were negatively correlated with memory test performance [68]. Moreover, the data of MRI and histological analysis suggest that increased iron load in the brain is also associated with Aβ plaque formation, which locally binds to the Aβ core and halo regions and is entangled with hyperphosphorylated tau (pTau) in humans and animals [107, 182, 260]. At the cellular level, oligodendrocytes require iron as a cofactor for enzymes involved in myelin production and maintenance and have the highest iron content of all cell types in CNS, which increases with age and even further in AD [21]. Compared to other cells, oligodendrocytes contain low levels of anti-oxidants [273], which renders them vulnerable to oxidative stress that can accompany a rapid increase in intracellular iron [199]. Thus, the observation that neuroimaging-defined white matter abnormalities are characteristic of AD is relatively new, which may be radiological manifestations in preclinical AD [85, 199]. In addition, the successful application of iron chelators in patients with early AD and several AD animal models [62, 78, 115, 114, 116, 142, 237, 309] provided strong evidence that iron-overloading in specific brain areas with age is associated with the pathophysiological progression of AD. Of note, an important consideration in studying iron levels in AD is to emphasize that the increase in total iron content in the brain is not necessary for brain oxidative stress alone, but that the body may favor oxidation of Aβ and NFTs by inducing an iron-rich environment, causing agitation and cell death, while other brain regions may suffer from impaired nerve function due to iron deficiency [24].

5.4 The Role of Iron in the Pathogenesis of AD

Iron is a necessary element demanded for virtually all living organisms and complicates in numerous biological reactions such as electron transfer oxygen transport, oxidative metabolism, and gene regulation. In recent years, numerous studies have revealed that abnormally elevated brain iron is the cause of certain neurodegenerative diseases [160]. Nonetheless, why iron accumulates in these disorders and other important questions regarding their pathogenesis remain unanswered [188]. The abnormally increased iron ions in the brain of AD patients can interact with H_2O_2 to produce highly active hydroxyl radicals in the body. Hydroxyl radicals act on the proteins, nucleic acids, and cell membrane containing a large amount of unsaturated fatty acids and then cause cell damage, called oxidative stress damage [161]. Mitochondria are an energy processing plant in cells. It is also an important site for generating free radicals. Excessive iron destroys the mitochondrial respiratory chain, the destruction of any part of the respiratory chain causes the formation of free radicals, and then the biological macromolecules are damaged, eventually the cells death [8, 131]. Accumulated iron can also damage mitochondrial membrane potential, activate caspase-3-dependent apoptotic pathway leading to cells apoptosis, and ultimately lead to the occurrence of AD [164].

5.4.1 *Iron and APP/Tau in AD*

In AD, Fe accumulation has been observed in and around the Aβ, SPs and NFTs [260]. Cerebral iron deposition may be related to the metal-binding sites of Aβ and tau proteins and may be related to abnormal expression of various brain iron metabolism-related proteins in the brain of AD patients.

Interactions between iron and Aβ or APP

APP is expressed in most cell types, both brain-derived and other tissues, including blood cells [238]. In the amyloidogenic pathway, APP is a transmembrane metalloprotein [194] that is cleaved by β-secretase enzyme and γ-secretase enzyme to generate the neurotoxic 40–42-amino-acid prefibrillar amyloid [286]. During the non-amyloidogenic path of APP metabolism, α-secretase enzyme is the zinc metalloprotease that cleaves APP in the Aβ-peptide domain to release the 90 kDa ectodomain of APP. According to reports, we know that α-secretase activation produces a 90 kDa APP(s) and APP mRNA translation upregulation after interleukin (IL)-1-mediated stress response, ultimately resulting in secretion of the APP, is increased [19]. Some study findings showed that APP provides protection against catalyzed oxidative stress, for example, the lesions during the pathogenesis of AD or lesion and reperfusion after stroke [91].

Aggregation of Aβ in brain is the main pathological characteristic of AD, which is known to be triggered by metals such as iron. Aβ can directly bind iron through its His6, His13, His14, due to the universality of the binding of ferrous iron to Aβ, and the interaction between Aβ and iron is believed to be beneficial to smaller microenvironments such as the brain [32, 298]. Recently, a study uncovering the relationships between iron biochemistry and AD pathology in APP/PS1 mouse cortex revealed that iron is an integral part of the SP and also provided the evidence that Aβ-induced reduction of iron can be occurred in vivo [269]. The iron-aggregated Aβ produces substantial ROS which would partially be conducive to the neurotoxicity present within the iron-enriched regions around SPs [96, 166].

Iron metabolism can also be modulated by Aβ. The proteins required to regulate iron metabolism in brain are quite similar to that in periphery. The iron response proteins (IRPs) 1 and 2 bind to the iron regulatory elements (IREs) in the 3′ or 5′-untranslated region of an mRNA to regulate the iron regulatory proteins' expression. Iron can not only accelerate the Aβ deposition, but also regulate the Aβ production. The intracellular iron was reported to modulate the APP translation via targeting the IRE in the 5′-untranslated region of APP transcript; therefore, the increased concentration of intracellular iron translationally upregulates the APP expression [50, 238]. Interestingly, the APP was recently proved to be as an anchor for ferroportin (FPN) which stables the cell-surface expression of FPN, thereby promoting the efflux of iron [75]. Depletion of APP in neuronal cultures or in mouse models resulted in the retention of iron in cells [75], while exogenously add the APP or overexpression of APP rescues this phenomenon [287]. Based on these observations, the iron overload modulates the Aβ generation via regulating the expression of APP; in return, the APP regulates the intracellular iron homeostasis by stabling the cell-surface FPN to increase iron efflux.

Iron and Tau

Tauopathies are mainly characterized by NFTs, which compose of spiral fibers by consisting of hyperphosphorylated tau. Tau promotes microtubule stabilization and regulation, regulates axonal transport [71, 174], and may play an important role in neurotransmission and iron homeostasis through the trafficking of APP to the cell surface [158]. These behaviors may be interrupted following hyperphosphorylation of tau proteins, and iron has been clarified to mediate the modification in keeping with its redox state. Hyperphosphorylation of tau, one of the key steps toward aggregation, may be induced by iron via CDK5/P25 complex and GSK-3β [115]. In addition, Fe^{3+} may mediate nitration of tau, which prevents microtubule stabilization and is commonly found in NFTs and SPs [303]. The accumulation of tau in tangles also results in induction of heme oxygenase (HO) 1, an antioxidant that facilitates the release of the redox-energetic Fe^{2+}, which releases loose radicals to produce oxidative stress [294]. This in flip promotes hyperphosphorylation and aggregation of tau [152]. In addition, iron accumulation leads to loss of tau function in the cells, which prevents iron export through impaired transport of APP to the cell membrane, where it stabilizes ferroportin resulting in a vicious cycle of iron accumulation and tau pathology [299]. Other studies however suggest that iron chelator was reported to dissociate the binding of iron associated with hyperphosphorylated tau [258], which might indicate better therapeutic strategies to suppress tau phosphorylation in AD [115].

5.4.2 Iron and Oxidative Stress in AD

Iron plays an important role in the brain. On the one hand, it maintains the brain's extremely high respiratory activity for the benefit of myelination and the production of various neurotransmitters [37]. On the other hand, due to the high oxygen metabolism rate of brain tissue and metal ions such as iron, and the relative lack of antioxidant protection mechanism, it is easy to cause a loss of balance between oxidative damage and antioxidant defense of brain tissue [49]. Oxidative stress is a common pathological feature inside the development and pathogenesis of several neurodegenerative diseases, which includes AD [161, 291]. Empirical evidence shows that oxidative stress occurs prior to the advent of SPs and NFTs, thus leading to the signs and symptoms of dementia [291]. Clinical data further prove an association between oxidative stress and AD-like cognitive dysfunction [140]. Oxidative stress contributes to the production of Aβ and hence aggravates its neurotoxicity [193].

In recent years, lots of studies have confirmed that substantially all intracellular macromolecules, such as lipid, protein and nucleic acid, existing in oxidized form in the brain tissue of AD patients, and the level of oxidized form are significantly increased.

Lipid peroxidation

It has been excellently revealed that lipid peroxidation comprises three distinct mechanisms, free radical-mediated oxidation, free radical-independent, nonenzymatic oxidation, and enzymatic oxidation [205, 250, 256]. Lipid peroxidation has been shown to induce disturbance to membrane organization and functional disorder of proteins and DNA. However, lipid peroxidation products also act as redox signaling mediators [311].

The aggregation of Aβ is recommended to play a central role in the pathogenesis of AD which is the most common cause of dementia, Aβ, and small oligomers can insert into the lipid bilayer and generate hydrogen peroxide [135]. In the presence of redox-active transition metal ions, consisting of iron and copper, hydrogen peroxide undergoes a Fenton reaction resulting in the formation of hydroxyl radicals. The reaction then develops lipid peroxidation and causes oxidative impairment to proteins and biomolecules [40]. On the other hand, using an animal model of AD, it has been shown that oxidative stress and lipid peroxidation can induce Aβ accumulation during the progression of AD [223]. The growing body of evidence for the involvement of lipid peroxidation and general oxidative stress as a driving force of AD has generated much interest in using antioxidants as therapeutics [94]. Since the antioxidant capacity of a cell is dependent on a cocktail of multiple antioxidants, it is possible that supplementation with one or two antioxidants is not sufficient for changing the course of disease in humans. Additionally, since lipid peroxidation can be initiated at very early stages of the disease, antioxidant therapy may be more effective in patients with early stages of AD, rather than late-stage patients [218].

DNA/RNA oxidation and accumulation

In the foregoing text of the present article, we have already mentioned that the brain is susceptible to oxidative stress and ROS caused by the presence of iron. Homeostasis of iron is tightly regulated under normal physiological conditions as excessive amounts of iron are known to cause cellular susceptibility to oxidative stress excess, and then injury DNA and/or RNA via ROS mechanism [128]. The accumulation of DNA and RNA oxidative damage is observed in cortical and hippocampal neurons from AD brains at early stages of pathology [283]. An increase in oxidative DNA and RNA damage occurs in a subset of cortical and hippocampal neurons at the onset of AD [33]. Non-repaired nucleic acid damage is highly deleterious in neurons, as it can trigger transcriptional or translational dysregulation and chromosomal instability, as observed in AD [58]. However, it is poorly comprehended about the mechanisms underlying the alteration of nucleic acid integrity in neurons all through the early tiers of AD pathogenesis [128]. In unique, the capability role of tau pathology in nucleic acid integrity has not been explored.

Several groups have documented the accumulation of oxidative DNA damage during aging in the brain [121, 127]. Oxidative damage to DNA may be one of the maximum crucial elements in neuron degeneration of AD due to the critical role of DNA in cell function. Lu et al. [171] located that oxidative DNA damage preferentially accumulates in the promoter regions of several genes concerned in synaptic plasticity, vesicular transport, and mitochondrial function. It has been mentioned that differential accumulation of oxidative DNA damage in the brain regions indicates that certain regions of the brain are more susceptible to oxidative DNA modifications than other regions [42, 101]. Without efficient repair ability in the brain in AD, mutations in nDNA and mtDNA may result in neuron death through defects in oxidative phosphorylation [93, 170]. Furthermore, higher concentrations of oxidized pyridines and purines were detected in lymphocytes and leukocytes of AD patients compared with controls [185, 190], and increased levels of 8-oxoG were observed in DNA from ventricular CSF of AD patients [170]. Recent findings indicate increased oxidative damage in leukocytes [190] and brain tissue [288] of subjects with MCI, indicating that DNA oxidation may constitute an early occasion in the AD progression, prior to cytopathological changes, and might make contributions to the pathogenesis of the disease [184].

Abundant evidence considerably supports the early involvement of RNA oxidation in the pathological of neurodegeneration, particularly in AD. RNA oxidation was observed in postmortem brains of early AD cases [208], a presymptomatic case with familial AD mutation [207]. Oxidized RNAs may demote faster than other types of oxidatively modified macromolecules, while protein carbonyl compounds, lipid peroxidation products, and glycoxidation products are stable at the point of production [210]. Researches proved the observations contrast remarkably with the pattern of the RNA oxidation, a steady-state marker, which is prominent in neurons without pathology [208, 209].

As for RNA oxidization, a study has shown that the Fenton reaction oxidized and cleaved rRNA in cultured primary neurons and astrocytes [69]. It is known that redox-active iron has been shown to also bind with RNA at the metal-binding site [25] and that redox-active iron bound to ribosomes oxidizes rRNA in AD [130]. Of course, mRNA oxidation is not a negligible event and may play an important role in the pathogenesis of AD [255].

Mitochondrial dysfunction

Mitochondrial dysfunction has long been associated with several neurodegenerative diseases including AD [278]. Mitochondria are essential for regulating the equilibrium of iron throughout the cell [219]. Mitochondrial function is critical for normal cellular processes. It is particularly notable that many neurodegenerative diseases are directly related to iron metabolism disorders in mitochondria [278, 293]. For example, an abnormal increase in iron in brain tissue of patients with AD and PD is likely to be one of the important causes of neuronal death [20]. Mitochondrial damage and oxidative stress have been greatly implicated in the progression of aging, along with the pathogenesis of age-related neurodegenerative diseases, such as AD, PD, and ALS [79]. Several mitochondria-targeted strategies, such as promoting antioxidant bioavailability through novel delivery systems, figuring out unique mitochondrial proteins as precise drug targets, studying signaling pathways for mitochondrial biogenesis and kinetics, and identifying effective natural products may be effective resistant to neurodegenerative diseases associated with mitochondrial dysfunction [268].

In fact, the early stage of AD is identified via morphologically and functionally impaired mitochondria, which are prone to over-form ROS, resulting in reduced brain energy due to ATP reduction [102]. Once formed, Aβ corresponds with the mitochondria and causes further mitochondrial defect. Aβ inhibits complex IV and binds Aβ-bound alcohol dehydrogenase (ABAD). ABAD produces ROS. It has been reported that mitochondria contain an active γ-secretase complex that is involved in processing of APP to form Aβ, and the non-soluble APP can affect outer mitochondrial membrane and interfere with protein shuttle import [102]. Impaired crosstalk between endoplasmic reticulum and mitochondria also seems to play an important role in AD pathogenesis. In addition, mitochondrial dysfunction and apoptosis are associated with Aβ-mediated toxicity [251]. In brief, mitochondria function as both the source and target of toxic ROS because mitochondrial dysfunction and oxidative stress are important in age-related neurodegenerative diseases, particularly AD [281]. However, whether mitochondrial iron accumulation is a founding event or a consequence of upstream events in a particular disease is a matter of debate [187].

5.4.3 *Neuroinflammation*

Neuroinflammation plays an important role in the pathogenesis of many of the neurodegenerative diseases. Increasing evidence suggests that the pathogenesis of AD may be caused by inappropriate activation of the inflammatory response in the brain, and a strong immune response may cause cell damage or death [124, 126]. In AD neuroinflammation, Aβ deposition alone may be sufficient to trigger an inflammatory reaction by binding to pattern recognition receptors on micro- and astroglia, characterized by the release of inflammatory mediators, which subsequently contributes to cognitive decline and disease progression. Given the possibility that Aβ deposition precedes cognitive deficits, one may speculate that exogenous or endogenous factors, including systemic inflammation, obesity, iron accumulation, and traumatic brain injury, can modify the innate immune response mounted by Aβ-exposed microglia [124, 183].

Indeed, both animal studies and clinical AD patients demonstrate SPs and NFTs formation in vicinity to the activated glial cells in the brain, and thus, immune response is aggravated [24, 129, 76]. The microglia and astrocyte activation stimulated by Aβ deposition induce reactive gliosis and pro-inflammatory signaling cascade [145]. Activation of astrocytes patients is often secondary to activation of microglia in the brain of AD, and activation of astrocytes enhances β-amyloid deposition by overexpression of APP, further activation microglia to form a vicious circle [245]. A renowned mark of neuroinflammation is the activation and increased intake of extracellular iron and subsequent downregulation of iron-interacting proteins, causing the intracellular iron sequestration [272]. Alter of morphology and quantity of microglia and astrocyte cells are observed in the AD brain, and both are involved in iron homeostasis [183]. Microglia isolates excess iron in ferritin to protect neurons. The incidence of L-ferritin positive active microglia is notably increased in the AD brain, and the characterization of L-ferritin positive cells showed that they are merely microglia in accordance with morphological and immunohistochemical staining [54, 169]. Microglial cells interact with both non-Tf-bound iron and Tf-bound iron [92, 206], and pathways for each transport substrate have been characterized [180].

Elevated brain concentrations of inflammatory cytokines have been associated with AD [245, 310]. Interestingly, accumulation of intracellular iron is associated with neuronal degeneration that underlies most neurological disorders [302], and microglial secretion of the inflammatory cytokines TNF-α and IL-1β enhances neuronal iron uptake [290]. In turn, these pro-inflammatory mediators have been shown to strongly influence microglia iron transport and metabolism [129, 144, 272, 277]. In addition, research findings pointed lipopolysaccharide (LPS)-induced expression of hepcidin in astrocytes is regulated by IL-6-activated microglia. Astrocyte-derived hepcidin is tightly related to neurons' apoptosis by iron accumulation in the neurons. These data indicate that a few iron-related proteins, including hepcidin, are not only a factor in the regulation of iron metabolism but also a critical linker between iron homeostasis and the modulation of acute inflammatory responses therefore further elaborate the pathogenesis of neurodegenerative diseases [306]. Despite the connection between inflammation and iron metabolism has been established, the relationship and mechanisms between neuroinflammation and iron metabolism in neurodegenerative diseases are not fully understood.

5.4.4 *Iron Affected NMDAR-Mediated Synaptic Transmission in AD*

NMDARs, a cation channel, adjust the entry of Na^+ and Ca^{2+} ions when are activated by the co-agonists glutamate (or NMDA) and glycine or D-serine. Appropriate activation of NMDARs plays a critical role in physiological functions such as excitatory neurotransmission, synaptic plasticity [45]. Activated NMDARs cause speedy opening of Ca^{2+} channels on the cell membrane, followed by an increase in free Ca^{2+} ions in the cytoplasm with the aid of a CNS signaling cascade [215]. However, excitotoxicity induced by excessive activation of NMDARs contributes to pathological changes in the CNS [106]. Hippocampal neurons require iron to generate calcium signals after NMDARs' stimulation [196]. Studies of NMDARs involvement in the NTBI influx pathways implied that activation of NMDARs drastically promotes Fe^{2+} entry into cells in primary neurons [48, 216]. Moreover, iron has been shown to block the influx of Ca^{2+} in the NMDARs channel, suggesting that Fe^{2+} competes with Ca^{2+} for NMDARs in cultured neurons [197]. In addition, NMDAR's activation has been mentioned to promote iron accumulation by accelerating DMT1-mediated iron influx [48], amplifying iron releasing from lysosome and regulating DMT1 expression [296]. Interestingly, it is the increased expression of DMT1+IRE mRNA rather than DMT1-IRE mRNA after treatment of NMDA in primary hippocampal neurons [118]. In fact, iron deficiency can inhibit cognitive ability [43, 81, 196]; thus, it is of potential value to ensure an adequate iron supply through the pathway of iron entry from DMT1. This means that NMDAR's activation may contribute to the formation of hippocampal spatial memory by stimulating the expression of DMT1+IRE.

Of note, a role for iron in NMDAR-mediated synaptic plasticity has begun to remind. A previous study suggest that ROS generated by iron are required for NMDA-induced and RyR-mediated ERK1/2 phosphorylation/nuclear translocation (a requisite step of sustained synaptic plasticity) in primary hippocampal neurons, and electrophysiological studies also found that iron addition resulted in a significant increase of the long-term potentiation (LTP), but the increase can be prevented by iron chelation [196]. Recent studies reported that NMDARs were substantially improved within the prefrontal cortex and hippocampus of iron-loaded rats, suggesting that altered iron status by iron supplementation in the brain regulates glutamatergic neurotransmission pathway [122]. In the hippocampus, studies found that intracellular iron signaling could modulate neuronal excitability [296], suggesting that iron can make in a feedback method to modulate NMDA function to maintain neuronal excitability under normal condition. Further study showed that iron-induced ROS generation is required for Ca^{2+} release for long-term potentiation in primary hippocampal neurons [196]. However, iron overload might promote NMDARs mediated excitotoxicity possibly by enhancing Ca^{2+} release and calcium-induced cell damage. Both iron dyshomeostasis and NMDAR-mediated neurotoxicity have been shown to have an important role in neurological diseases such as AD [301].

5.4.5 Iron and Neuronal Damage

Synaptic loss and neuronal death are one of the major pathological hallmarks of AD, and the endurance of neuronal loss is correlated with the progression of the disease. Synapses are formed by the junction between two neurons, which allows neuronal cells to transmit signals to another cell. This channel is usually damaged or lost in many neurodegenerative diseases, such as AD [143, 257]. Accumulating evidence suggests that dysfunction and loss of synaptic connections may be an important early event underlying AD progression, more than other pathological changes [12, 57, 262]. Damage to synapses by Aβ oligomers includes changes in synaptic numbers, synaptic transmission, and changes in synaptic plasticity [80, 224, 262]. Comparably, evidence also shows that pathological changes in tau are toxic for synapses [222]. The molecular mechanisms that lead to synapse degeneration and neuronal loss downstream of Aβ and tau are less well established, but different pathways are implicated such as metal dyshomeostasis, oxidative stress, mitochondrial dysfunction, and neuroinflammation [64]. Some results suggested that iron impacts mitochondrial dynamics, possibly trigging synaptic loss and apoptotic, and suggests that iron chelators have to take into consideration as a capability molecule with memory restoring and neuroprotective properties, observed in the remedy of cognitive deficits in neurodegenerative disorders [259, 117]. Reduced synaptophysin may be associated with iron-induced cognitive deficits observed in AD model, as these researchers have previously reported that iron-induced hippocampal synaptophysin levels continue to decline in rats [259, 155]. Recently, some data on iron deficit that may play a role in synaptic processes, as well as preceding studies on the effects of iron deficiency on cognitive processes and synaptic function, can generate valuable information to prevent against the effects of iron deficiency and iron overload [195]. Thus, insightful synapse degeneration in AD is characterized by the worsening of cognitive function, synapse loss, and neuronal cell death.

Morphologically, advanced AD manifests postmortem as a significant shrinkage in cortical volume, especially throughout the cortex and hippocampus. This is caused by massive neuronal cell death, whose etiology is not clear [88]. Notably, the type of cell death that occurs in affected areas of the AD brain is still contentious, despite the demonstration that DNA fragmentation and upregulation of pro-apoptotic proteins has been frequently observed [227]. As such, it remains unclear to what extent apoptosis or emerging types of regulated necrosis are responsible for bulk neuronal loss in AD [16, 227]. We now know that abnormal elevation of iron may be the initiating factor of neuronal death in the brain of patients with neurodegenerative diseases [225]. Iron-induced oxidative stress has been shown to initiate several apoptotic signaling pathways in neurons [243] and cause oxidative damage to key proteins. There is extensive evidence that oxidative damage to proteins and lipids by iron can cause synaptic dysfunction and neuronal cell death [178], both of which are critical features of AD.

Ferroptosis is the fourth cell death pattern proposed by scholars based on the unique pathological state caused by intracellular iron accumulation, which is different from necrosis, apoptosis, and autophagy [73]. Ferroptosis signs a newly located form of cell death that is not associated with the caspase pathway and is involved in iron disorders, lipid peroxidation, and inflammation as predominant markers [72]. The disorder in cellular redox homeostasis leads to accumulation of lipid ROS and hence lipid peroxidation [1, 97]. Morphological changes affecting the outcomes involve mitochondrial volume reduction, reduced density of mitochondrial membranes, chromatin condensation, eventually rupture of mitochondrial outer membrane, and plasma membrane [73, 89]. Several powerful ferroptosis inducers have been identified [267] in which the small-molecule erastin effectively inhibits xCTcystine back transporter to further induce ferroptosis, as cystine is vital for glutathione synthesis. Glutathione peroxidase 4 (GPX4) [305] utilizes reduced glutathione to detoxify phospholipid hydroperoxides and hydrogen peroxides [279]. It should be noted that GPX4 knockout indicates iron imbalance, lipid peroxidation, and inflammation in mice, at the same time shows early signs of AD phenotype, including behavior defect and hippocampal neurodegenerative disease [120, 253]. Intriguingly, small-molecule inhibitors of ferroptosis can mitigate neurodegeneration and inflammation observed in GPX4 knockout mice, suggesting that ferroptosis is a potential neurodegenerative mechanism affecting neurons important for spatial function [120].

5.4.6 Iron and White Matter

Over the past several years, neuroimaging studies have implicated micro- and macrostructural abnormalities in white matter in the risk and progression of AD [34, 35, 157]. Analysis of postmortem brain tissue of AD patients found out a chemical change in white matter compared to non-demented patients: The amounts of general protein, myelin basic protein, myelin proteolipid protein, cyclic nucleotide phosphohydrolase, and cholesterol are notably reduced, indicating myelin loss [239]. The abnormal white matter in AD is mainly manifested in changes in myelin and oligodendrocytes [41, 199]. Myelin loss and the inability of the oligodendrocytes, the cells responsible for the production and maintenance of myelin, to repair myelin damage may be additional central features of AD [41]. Due to the essential role of oligodendrocyte lineage in the process of myelination and remyelination, changes in the variety of oligodendrocytes or their precursor cells and/or their dysfunction can affect myelin integrity and thus may be related to AD pathogenesis [179].

In the brain, iron is a key substance involved in the process of myelin protein and fatty acid synthesis in oligodendrocytes. Iron deposition in the brain of AD patients, which in turn causes an increase in oxidative stress, may result in increased cell damage. It is reported that myelinating oligodendrocytes contains a lower amount of antioxidant than other glial cells [141]. Therefore, high iron content and low antioxidant content make oligodendrocytes one of the most susceptible to oxidative stress in the CNS [276]. Rather, iron deficiency during infancy can lead to neurocognitive developmental delay, impaired learning and memory and in some cases can lead to mental illness, in part due to impaired myelination [228]. Due to normal aging, loss of white matter increases free iron, which is toxic to the brain. MRI analysis showed that white matter damage in AD patients preceded gray matter damage compared with age-matched controls [3].

5.5 Iron-Binding Proteins Implicated in AD

Ferritin

Ferritin is a highly conserved supramolecular protein, which composes of two different subunit types, H (heavy) and L (light). These two subunits co-assemble into a 24-subunit heteropolymer, with tissue-specific distributions, to form shell-like protein. A single ferritin molecule can sequester up to thousands of iron atoms, in the form of insoluble inorganic ferrihydrite core [14]. As a ubiquitous and well-characterized iron storage and detoxification protein, ferritin is the most efficient way to protect cells from the toxic effect of elevated levels of labile iron cations. In bacteria, plants, fish, as well as amphibians, the ferritins are generally homopolymers which consist of H-type subunits, while the ferritin is a heteropolymer typically composing of 24 subunits of two types, H and L in animals [14, 186]. The inner cavity of ferritin is separated from the outside environment by the resultant protein nanocage which is remarkably stable in the pH range of 3–9 and temperatures up to 80 °C. Small molecules and various cations can be able to enter the ferritin interior through eight narrow hydrophilic threefold channels also maybe through six hydrophobic fourfold channels as well [14]. The natural Fe(II) cations rapidly diffuse into the ferritin shell, where the Fe(II) cations are oxidized to insoluble iron(III) cations to form the inorganic ferrihydrite core. The different subunits of ferritin own different bio-functions. In mammals, the H-subunit is charging of the rapid oxidation of Fe(II) to Fe(III) by oxidative molecules at a dinuclear center, whereas the L-subunit helps iron export from the ferritin interior in support of iron nucleation and mineralization [31, 270]. It should also be noted that ferritin consisting exclusively of L-subunits is still able to oxidative Fe(II) cations, although this bio-activity is very low. It was reported that ferritin in plasma and cerebrospinal fluid is significantly increased in neocortical amyloid-β loaded participants, and ferritin is also observed to correlate positively with amyloid-β load[15, 109]. Subsequent studies implicate that CSF ferritin levels were able to act as a predictive biomarker of disease progression in the prodromal stages of AD [70].

Mitochondrial ferritin

Mitochondrial ferritin (FtMt), specifically located in mitochondria, is a new ferritin type identified in 2001. The human FtMt gene locates at chromosome 5q23.1 which encodes a ~1 kb mRNA that translates to a 242 amino acids FtMt precursor protein. The FtMt precursor protein is translocated to mitochondria after synthesis and is further processed to the mature type as the subunit to form typical ferritin shells [61]. The mature human FtMt has a 79% identity to the H-chain ferritin in the sequence of amino acids. The inner cavity of FtMt and H-ferritin shares a highly conserved sequence and a fully overlapped crystallographic structure [153], suggesting that these two proteins have similar functions. Unlike ferritin, the expression of FtMt is tissue-specific, and it is predominantly expressed in the testis and brain [244]. The major bio-function of FtMt is also sequestering excess free iron. Although the ferroxidase activity of FtMt is not as strong as H-ferritin, the iron sequestering ability is high. Overexpression of FtMt in Hela cells resulted in reduced expression of cytosolic ferritin and cytosolic iron deficiency [61]. Elevated expression of FtMt was observed in the brains of AD patients and was associated with anti-oxidative activity [304]. Furthermore, FtMt deletion exacerbated Aβ-induced neurotoxicity [292], whereas FtMt overexpression attenuated Aβ-induced neurotoxicity [300].

Lactoferrin

In 1939, the lactoferrin (Lf) was firstly identified in bovine milk and subsequently isolated in 1960 from both human and bovine milk. Human Lf is a glycoprotein which consists of 691 amino acids, which is constitutively expressed and secreted by glandular epithelial cells and by neutrophils following induction. The Lf possesses multi-functional bioactivities, such as anti-microbe, anti-inflammation, and immunomodulation; meanwhile, it is an iron-binding glycoprotein, which is important for the modulation of iron homeostasis. Like Tf, Lf has been shown to reversibly bind iron as Fe^{3+} with high affinity ($K_D \sim 10^{-22}$ M), as well as retain Fe^{3+} until pH values less than 3.0. The Fe^{3+}-binding activity of Lf is dependent on the concomitant and synergistic binding of carbonate, CO_3^{2-} [5]. Spectroscopic studies and the 3D structure suggest that Lf binds the CO_3^{2-} firstly, thus neutralizing the positive charge of the arginine residue (Arg-121 in the N-lobe and Arg-465 in the C-lobe) [18]. The participation of the CO_3^{2-} in the iron binding to Lf is important for iron reversible binding since the protonation of the CO_3^{2-} is likely a first step to break up the iron site at low pH values [2]. Lf has two main conformational states, the open metal-free (apo-lactoferrin, apo-Lf) state and the closed metal-bound (holo-lactoferrin, holo-Lf) state. Metal binding and release are determined by the large-scale conformational changes of Lf, by which the domains open to release the irons or close over the bound irons [99]. The metal-free apo-Lf form is flexible and more prone to thermal denaturation and digestion by proteases. However, the metal-bound holo-Lf is conformationally rigid and very stable, and the bound irons can only be removed by the very low pH or some iron-chelating reagents. Of note, the iron-binding activity of Lf is not specific, and other trivalent metal ions such as Ga^{3+} and Al^{3+} can also bind to Lf, although the affinity is lesser than Fe^{3+} [17]. The functions of Lf in AD pathology remain largely unknown. A recent study reported that intranasal human Lf effectively inhibits the cognitive decline of APP/PS1 mice via activation of ADAM10 expression [116]. Meanwhile, salivary Lf was identified as a diagnostic biomarker for mild cognitive impairment and AD due to it is more accurate than the cerebrospinal fluid total tau and Aβ42 [44].

Hepcidin

In 2000, a defensin-like disulfide-bonded 25 amino acid peptide with antimicrobial activity was respectively identified by two independent research groups in human ultrafiltrate and urine [151, 214]. This peptide was named hepcidin, and structural study further uncovered that this peptide contains eight disulfide-bonded cysteines forming a structure of β-turns, loops, and distorted β-sheets [214]. Hepcidin is predominantly synthesized and secreted by hepatocytes. In iron-overloaded mice, a dose-dependent induction of hepcidin mRNA transcript was observed [220]. However, substantial irons were accumulated in liver, pancreas, and serum, while the splenic iron store was depleted, in the mice with lack of hepcidin gene expression [202]. These studies revealed that hepcidin plays a vital role in regulating homeostasis of body iron. Upon binding to FPN, hepcidin induces covalent modification of FPN and subsequently causes FPN internalization and degradation in lysosomes, which reduces the export of iron from reticuloendothelial macrophages, hepatocytes, as well as duodenal enterocytes [201]. Hepcidin decreases the cell face stability of FPN, thereby inhibiting the entry of iron into plasma. Decreasing iron efflux from cells coupled with the ongoing iron uptake of erythropoiesis through the bone marrow leads to decreased serum iron and reduced transferrin (TF) saturation. Agelong high levels of hepcidin lead to insufficient serum iron for erythropoiesis which increases the risk of iron-restricted anemias [203]. In AD brains, hepcidin expression was decreased and restricted to the blood vessels, neuropil and damaged neurons; moreover, hepcidin was accompanied with heme-positive granular deposits in the region of damaged blood vessels [229]. The reasons for the reduction in hepcidin levels are less clear which need to be further investigated.

Ferroportin

FPN, encoded by the SLC40A1 gene, a 571 amino acid transmembrane protein, plays a critical role in iron metabolism. Although cells possess several mechanisms to uptake iron, FPN is the only known cellular iron exporter [125]. Iron efflux is in the form of ferrous iron; therefore, the activity of ferroxidase is necessary for FPN. The ferroxidase converts the divalent iron into ferric iron, which is mainly accomplished by the positive membrane iron transport accessory proteins or ceruloplasmin. Fpn controls iron entry into the plasma through duodenal enterocytes (1–2 mg/day) or macrophages that recycle near 20–30 mg/day of iron from senescent erythrocytes [74]. Recently, FPN has been recognized as the receptor for hepcidin, a master regulator of iron homeostasis [235]. Upon binding to hepcidin, FPN is internalized and subsequently submitted to lysosomal degradation [201], which results in a reduction of plasma iron concentration and an increase of intracellular ferritin. The FPN expression in AD brain is reduced with uncertain reasons [229]. Considering that APP stabilizes the cell-surface FPN expression [75] and substantial APP variants were observed in AD patients, the decreased FPN expression may be highly associated with APP mutations.

Transferrin and transferrin receptor

TF–transferrin receptor (TFR) axon is an important pathway for iron homeostasis. All vertebrates have functional TF, which is a 76 KD glycoprotein that mainly expressed in and secreted by the liver with a half-life of approximately 8 days in serum. TF contains two highly homological lobes at the N- and C-termini, and these lobes are connected by a short linking region [104]. Each lobe can bind one metal ion such as zinc, gallium, and iron with different affinities; therefore, one TF can maximum bind two ferric Fe. Most iron in the serum is bound to the TF; with the blood circulations, the iron can be transported to every tissue and organ. Serum TF can exist in non-bound (apo-TF), monoferric, or differric (holo-TF) forms. TF-bound iron enters cells which needs TFR, and TFR is a 97 KD type 2 membrane protein expressed in as a homodimer in cell membrane. On the cell surface, holo-TF binds to TFR and subsequently initiates the endocytic process to intake iron [156]. The interaction of TF and TFR is determined by the pH value: at pH 7.4, and TFR binds to holo-TF but not apo-TF; inversely, at a lower pH value in endosomes, TFR binds to apo-TF [285]. In the acidic endosomes, the ferric Fe is dissociated from TF, and it is further converted to ferrous Fe by metalloreductases [211]. The TF-TFR complex in the endosomes is then transferred to the cell surface by recycling endosomes and subsequently release of apo-TF [74]. In AD plasma, the iron is decreased due to TF desaturation [123]. Furthermore, the phosphorylated TF in brain and serum from an AD animal model are significantly reduced, and the cerebrospinal fluid and serum from AD patients also exert an aberrant phosphorylated TF [241].

DMT1

DMT1 is the firstly identified iron transporter in mammalian, and it transports various ferrous metal ions including Mn^{2+}, Co^{2+}, and Fe^{2+} [138]. DMT1 has 12 transmembrane helices with aqueous cavity and a metal ion binding site formed between residues in α-helices 1 and 6. DMT1 is well known to play a vital role in iron metabolism. In the proximal small intestine, DMT1 couples with the transport of H and Fe^{2+} into enterocytes. Dietary iron in the small intestine is in the form of Fe^{3+}; therefore, a reduction step mediated by the apical reductase Dcytb is required prior to the uptake of iron by DMT1 [181]. DMT1 is abundantly expressed in brain, in aged rat brain; the expression of DMT1 is increased in an age-dependent manner which associated with elevated brain iron content [172]. This study indicates that age-induced DMT1 upregulation is associated with increased brain iron in AD brain.

Melanotransferrin

Melanotransferrin (MTf) is a TF homolog that is found mainly bound to the cell membrane via a glycosylphosphatidylinositol anchor. It shares an identical N-terminal Fe-binding site with serum TF, by which, MTf can bind one Fe-atom [36]. MTf can be secreted by cells at a very low level; however, its ability to donate Fe to cells is far less effective than TF [87]. Furthermore, the secreted MTf cannot be bound by the transferrin receptors, despite its high structural homology to TF [87]. Substantial in vitro studies have been performed to uncover the relationship between MTf and iron metabolism. However, the in vitro data all suggested that MTf does not play a vital role in Fe transport and metabolism [149, 234]. In addition, the MTf knockout mice showed normal fertility and development, with no morphological or histological abnormalities [77, 254]. Taken together, although the high-affinity Fe-binding site of MTf is structurally and functionally similar to that in Tf, MTf is not necessary in Fe transport and metabolism. It has been reported that MTf is expressed in reactive microglia and the secreted MTf is increased in sera in AD patients [148]. To date, the function of MTf in AD is largely unknown, which need to be further elucidated.

HFE

HFE is a protein that comprised of 343 amino acids, which interacts with β2 M to form a heterodimer expressed at the cell surface [83]. HFE contains a signal peptide, an extracellular transferrin receptor-binding region (α1 and α2), an immunoglobulin-like domain (α3), a transmembrane region, and a short cytoplasmic tail [83]. Glycosylation of HFE is important for its normal intracellular trafficking and functions. HFE can be glycosylated at residues N110, N130, and N234 during transport to the cell membrane [26]. HFE plays an important part in the regulation of iron homeostasis. In iron-loaded HFE knockout mice, liver hepcidin was significantly decreased [4]. Furthermore, iron overload was rescued in HFE knockout mice with constitutive overexpression of hepcidin [204]. These data suggested that HFE acts as an upstream regulator of hepcidin expression. HFE can interact with TFR1 and TFR2 in the cell membrane, and any changes in the expression of either HFE or TFR2 can cause the downregulation of hepcidin [111, 146]. The binding affinity of HFE for TFR is similar to the binding affinity of iron-loaded TF, and the HFE can competitively bind to the iron-loaded TF-binding site of TFR, thereby inhibiting the iron uptake [84]. The associations between HFE polymorphisms and AD were investigated in some studies. The HFE H63D mutation is associated with prolonged endoplasmic reticulum stress and increased the risk of AD [165, 175]. The clear functions of HFE in AD pathology remain unexplored, possibly involved in cholesterol modulation and iron regulation [6].

HO-1

HO-1 is one of the three isoforms of heme oxygenase (HO) family which act as key rate-limiting enzyme in the degradation of iron-containing heme [173]. It is widely distributed in many mammalian tissue cells and is also recognized as heat shock protein 32 (Hsp32). The level of HO-1 is low at basic conditions and can be highly increased in the response of numerous pro-oxidant agents and stimuli including heat shock, heavy metals, ultraviolet rays and cytokines. After induction, HO-1 plays a pleiotropic, protective role and shows antioxidant, anti-apoptotic, and anti-inflammatory effects in various conditions [168]. The cytoprotective effects of HO-1 depend on multiple cellular processes including the generation of bilirubin, an antioxidant molecule, from the degradation of heme; the induction of ferritin, a strong chelator of free molecular iron; and the liberation of CO, responsible for HO-1 major anti-inflammatory and anti-apoptotic effects [274]. Ninety percent of the body's iron is derived from the iron of aging red blood cells, which is the iron produced by the decomposition of hemoglobin. Therefore, the role of HO-1 in iron metabolism is crucial. Studies showed that sustained HO-1 activity decreased the expression of transferrin receptor-1 and increased the proteasomal degradation of ferritin [162]. Because of the distinct transcriptional regulation hemoglobin have from iron, the change of HO-1 also affect FPN1 and hepcidin [177]. Besides, HO-1 is also related to neurodegeneration diseases [247]. The expression levels of HO-1 in reactive astrocytes are markedly increased in the hippocampus and cerebral cortex of patients with AD relative to age-matched, non-demented controls [249]. In AD mice, long-term overexpression of HO-1 causes phosphorylation of tau protein, which accumulates in the brain and forms neuronal fiber tangles, aggravating the condition [137]. To review most of the literature, the glial HO-1 response may serve as a robust transducer of noxious stimuli, an important driver of relevant neuropathology and a potentially disease-modifying therapeutic target [248].

Lipocalin-2

Lipocalins are small secreted proteins that play roles in many processes including retinol transport and prostaglandin synthesis. The lipocalins are capable of binding low-molecular-weight ligands due to a common secondary and tertiary structure corresponds to a single, eight-stranded antiparallel β-barrel around a central pocket [46]. Lipocalin-2 (LCN2), also known as oncogene 24p3 or neutrophil gelatinase-associated lipocalin (NGAL), is a 25 kDa soluble and secretory protein and reportedly functions as a mediator of inflammation and iron ion transport [191]. Studies showed that LCN2 mediates iron ion transport by association with a siderophore. LCN2 saturated with iron (holo-form) can increase intracellular iron levels by moving and releasing iron into the cytoplasm. Iron-free form of LCN2 (apo-form) can deplete intracellular iron and transport it to the extracellular space via its receptor NGAL-R [66]. It can increase or decrease intracellular iron content based on the body iron store, making it a malleable factor in the maintenance of iron homoeostasis [47]. Of note, LCN2 protein levels in blood increase with age and mild cognitive impairment [51] and are increased in human postmortem brain tissues of AD [65, 200]. In the context of AD pathology, in vitro studies demonstrated that astrocytes generate high levels of LCN2 in response to Aβ and that LCN2 sensitizes brain cells more vulnerable to Aβ-induced cell death [189, 200]. However, more insight into such effects of LCN2 using LCN2 KO mice is required to understand its importance in AD and its potential as a therapeutic target [65].

Furin

Furin is a member of the mammalian proprotein convertases' family. It is a ubiquitously expressed calcium-dependent serine endoproteases that is expressed in all vertebrates and many invertebrates. It is expressed as a 794 amino acid zymogen that undergoes two autocatalytic cleavage steps before it becomes fully active [10]. As a proprotein convertase, furin is known to control the activation of diverse functional proteins by cleaving inactive protein precursors in the secretory pathway. Studies showed that hepcidin is produced in the liver and processed into the mature active peptide by furin. The hepcidin gene encodes a precursor prepropeptide of 84 amino acids. Furin recognizes the consensus sequence and participates the two sequential cleavages to produce the mature peptide of 25 amino acids [280]. Few studies point out that furin plays an important role in the activation of both α- and β-secretase. Both ADAM10 and ADAM17 (the α-secretase) contain a consensus furin motif which is essential for their activation. The β-secretase, known as BACE, also requires proteolytic removal of its proregion by furin at a minimal furin site [271].

5.6 Alteration of Iron Content for Clinical Diagnosis of AD Progression

The role of iron in the pathology of AD has only received attention in the past few decades, in part because of inconsistencies in the literature regarding whether iron increase is a major or minor event in AD brain [252]. Moreover, iron deficiency during infancy can lead to neurocognitive developmental delay, impaired learning, and memory and in some cases can lead to mental illness, in part due to impaired myelination [228]. However, an early inductively coupled plasma mass spectrometry (ICPMS) study showed that even in the normal brain region, there is a regionally specific change in metal content [230]. Thus, it is important to distinguish between large organizations and specific areas of research for the iron status in tissues of AD patients relative to age-matched controls, including the blood, cerebrospinal fluid (CSF), and brain tissue.

5.6.1 CSF Iron Level in AD

Using ICPMS technology, Gerhardsson et al. [98] detected CSF iron levels in 174 AD patients (122 females, 52 males) and 54 control subjects (36 females, 18 males). There was no significant difference in CSF iron levels between the two groups, and the same results were obtained by using the same method in different laboratories [30, 134]. Lavados et al. [154] used atomic absorption spectrometry (AAS) to detect iron levels in the CSF of 7 AD patients and 12 control groups. There was no significant difference in CSF iron levels between the two groups, consistent with the previous study [192]. Interestingly, in the above studies, there was no significant difference, but the CSF iron in the control group was slightly higher than the experimental group. Taken together, the iron content in the CSF of AD patients did not change significantly in its pathological process.

5.6.2 Blood Iron Level in AD

For iron levels in AD blood, the researchers used different experimental methods to verify the iron content in serum, plasma and whole blood of healthy controls and AD patients. Alimonti et al. [7] used inductively coupled plasma atomic emission spectrometry (ICPAES) to detect serum iron levels in 53 AD patients (36 females, 17 males) and 124 control subjects (43 females, 81 males). They found that the serum iron levels in the control group were higher than those in the AD group. Using ICPMS, several research teams also detected that healthy controls have higher serum iron levels than patients with AD [23, 27, 30]. However, Giambattistelli et al. [100] used a turbidity immunoassay kit to determined serum iron levels in 107 AD patients and 52 healthy controls, and there was no significant difference in serum iron levels between the two groups. Similar findings were detected by both Vitros 5.1 [275] and FerroZine colorimetry [264, 265], respectively. Using AAS, the observation of several research teams also confirmed that there was no difference in serum iron content between normal people and AD patients [136, 192, 263], while Ozcankaya et al. (Ozcankaya and Delibas [213]; Vural et al. [284]) detected that serum iron levels were significantly increased in AD patients than in the control group. Different conclusions from various studies may be due to different experimental methods, different data collection times, individual sample differences, and different geographical distribution of samples. Therefore, changes in iron levels in the blood may not be sufficient to diagnose AD. Compared with blood and CSF, changes in iron content in the brain have reference significance for the diagnosis of AD.

5.6.3 Brain Iron Level in AD

Iron has been extensively studied in different brain regions of AD patients, and the results are different. Corrigan et al. [60] used ICPMS to detect hippocampal iron levels in 12 AD patients (10 females, 2 males) and 12 control subjects (4 females, 8 males). The hippocampal iron levels in AD patients were significantly increased compared to the control group. Gu et al. [113] used the same method to detect cortical iron levels in 12 AD patients and 11 control groups and found that cortical iron levels were significantly increased compared to the control group. House et al. [133] used ICPAES to detect 5 AD patients (2 females, 3 males) and 5 control groups (2 females, 3 males). The iron levels of amygdala, corpus callosum, cingulate cortex, caudate nucleus, entorhinal cortex, frontal cortex, globuspallidus, hippocampus, parietal cortex, lateral temporal cortex, thalamus, subcortical frontal white matter, and temporal white matter found that except for lateral temporal cortex, iron levels were significantly increased compared to the control group, and the iron content in the other brain regions did not change significantly. Srivastava et al. [266] used ICPMS to detect iron content in the cerebellum and parietal cortex of 4 AD patients and 4 control groups and found no significant difference between the two groups. Galazka-Friedman et al.

[95] and Griffiths et al. [112] used AAS to detect iron content in hippocampus of AD patients and healthy controls and found that the iron content in hippocampus of AD patients increased slightly, but there was no significant difference. Andrási et al. [11] used instrumental neutron activation analysis (INAA) to detect iron content in the whole brain of 5 AD patients and 5 control groups. No significant difference was found between the two groups, but Cornett et al. [59] using the same method found that the iron content in multiple brain regions of AD patients was significantly increased. From the results of autopsy, iron is unevenly distributed in the brain of AD patients, and there are certain metabolic disorders, which may be involved in the pathological process of AD. Although the results are different, in general, iron has deposition in different brain regions of AD patients. This phenomenon may be one of the causes of memory loss in AD patients. Therefore, the increase of iron content in the brain may be one of the methods for the diagnosis of AD.

In fact, the diagnosis of AD is difficult because pathologic AD findings are not sufficient for symptoms. To investigate AD pathological development and biomarkers for early-stage AD identification, we learned that neuroimaging data do not directly measure AD-related pathology but suggest neurodegeneration and achieved many meaningful results; particularly, the noninvasive detectability of Aβ plaques in MRI has so far been largely attributed to focal iron deposition accompanying the plaques at preclinical stages of AD [105].

5.7 Targeting Iron Dysregulation

The issue of iron dysregulation has been the target of therapeutic attempts for the treatment of AD. In this regard, iron chelation is a potential therapeutic strategy in both oxidative stress and protein aggregation processes but remains a matter for debate. For a metal chelator to effectively exert chelation, it must have the following characteristics: (1) target iron-enriched areas without depleting transferrin-bound iron from plasma; (2) remove chelated iron from iron accumulation sites or transfer it to other biological proteins, such as circulating transferrin; (3) easy to penetrate the cell membrane and blood-brain barrier (BBB); and (4) no side effects or minor side effects on the body [28].

Several iron chelators including desferrioxamine (DFO) and clioquinol (CQ) have been recently designed for their potential use in the treatment of AD [242]. DFO, a hexadentate ligand with a higher binding affinity for Fe^{3+} than Fe^{2+}, is able to inhibit the formation of β-pleated sheets of Aβ(1–42) and dissolve preformed β-pleated sheets of plaque-like amyloid in vitro [132]. Despite a variety of promising results in some clinical trials [62], experimental mouse models, the molecular size, and hydrophilicity of DFO prevent it from freely crossing the BBB, when administered orally [231]. Again, long-term administrated DFO leads many side effects, including allergic reactions, anemia, liver and renal dysfunction, and neuronal hearing loss [29]. Our recent studies demonstrated that intranasal DFO treatment inhibits iron-induced amyloidogenic APP processing, tau hyperphosphorylation, and impaired spatial learning and memory [115, 114], suggesting that intranasal administration of DFO may be a safe and noninvasive therapeutic strategy for AD therapy. CQ, a small lipophilic iron chelator that can cross the BBB, has been used in phase II clinical trials for moderate AD cases [236]. However, CQ is not a iron selective and produces very toxic effects. Surprisingly, recent research implicating the role of CQ in the inhibition of cAMP efflux might represent promise for the treatment of neurodegenerative diseases [217]. In addition, plant polyphenols were proposed for the treatment of AD because of their antioxidant activities and strong iron-binding affinity [176], but their capacity to be absorbed at the gastrointestinal tract and be transported subsequently through the BBB is limited. Clearly, the development of an effective non-toxic therapeutic agent for AD-like brain disorders remains a challenging task.

α-lipoic acid (LA), a small-molecule compound that can be naturally synthesized in mammals, is mainly used for the treatment of diabetic polyneuropathy, and it was recently found that 129 patients with AD who had been treated with LA for a period of time had an effective remission of cognitive decline [82]. In fact, in vitro studies have demonstrated the antioxidant properties of LA and a similar role as metal chelating agents [212]. Studies showed that LA not only reduced iron uptake rates but also decreases dynamic iron pools by increasing iron stores within the lens epithelial cells [110]. In vivo, LA can enhance Aβ solubility the frontal cortex of APP transgenic mice and human AD brain tissues, suggesting that LA, like other transition metal chelators, reduces deposition of amyloid protein through Aβ re-dissolution [86]. Most recently, our study showed that chronic treatment with LA effectively inhibited tau hyperphosphorylation by alleviation of oxidative stress, inflammation, and ferroptosis in the brains of an AD mouse model [309].

Surprisingly, the recent study found that Lf as an iron chelator and a therapeutic agent has been investigated to identify its roles in AD. The affinity of Lf for iron is 300 times higher than that of TF. In weakly acidic environments, the affinity is further enhanced, which may be related to the transfer of iron from transferrin to Lf during inflammation [67]. Several experiments showed that Lf is expressed in many neurons and in some astrocytes, microglia, and oligodendrocytes of the human brain [13, 147] and that it could involve in aging processes [159, 226]. Further research findings suggested that Lf is highly expressed in both extracellular SPs composed of aggregated Aβ peptide and in intracellular NFTs composed of aggregated hyperphosphorylated tau protein, which are suggesting the link of Lf and AD-like pathologies [147, 289]. Therefore, the use of exogenous Lf as a therapeutic agent has been investigated to identify its roles in AD. Administrated Lf could stimulate the non-amyloidogenic processing of APP and α-secretase catalytic activity and expression in AD model mice and Lf enhanced the α-secretase-dependent APP process through the ERK1/2-CREB and HIF-1α pathways in vitro and in vivo [116]. In combination with other biological functions of Lf, such as its participation in metal ion metabolism and the regulation of cell proliferation and apoptosis, we infer that Lf may be a potential therapeutic agent to delay the pathological progress of AD.

5.8 Conclusion

Why is it so hard to find a cure to AD? The main explanation is that the complex causes of AD. It presents itself typically late in life and under the influence of many combined genetic and lifestyle risk modifiers such as metabolic disorder of metal ions, smoking, alcohol use, diabetes, body weight, hypertension, and physical and intellectual activity levels, and perhaps even certain diets. Researchers have done a lot of research on these pathogenic factors [103, 246, 282]. Recently, the study of the pathogenesis of AD using iron as a target has become a research hotspot [8, 32, 56, 160, 161, 164, 176, 242]. The use of iron chelators to chelate excess iron in the brains of AD patients has also become a new strategy for the treatment of AD [78]. However, many therapeutic drugs for AD are still difficult to translate into clinical application, largely due to the existence of BBB, which prevents drugs from entering the brain to play a role. Therefore, effective drugs for AD must be easy to pass through BBB. In our bodies, many proteins or molecules bind to iron to form active complexes, such as LA and Lf. Some researchers have confirmed that treating AD transgenic mice with LA and LF can alleviate their pathological symptoms [116, 309], but the specific mechanism of its action is still unclear, and further studies are needed. Hopefully, in the meantime, Lf-based nano-drug molecules are also under study. This drug molecule exhibits excellent biocompatibility and can easily cross BBB to reach the lesion site, showing effective therapeutic effect. Future research will also use iron as an opportunity to study the mechanism of the occurrence and development of AD from the iron homeostasis to more fully explain the causes of AD.

References

1. Abdalkader M, Lampinen R, Kanninen KM, Malm TM, Liddell JR (2018) Targeting Nrf2 to suppress ferroptosis and mitochondrial dysfunction in neurodegeneration. Front Neurosci 12:466
2. Adams TE, Mason AB, He QY, Halbrooks PJ, Briggs SK et al (2003) The position of arginine 124 controls the rate of iron release from the N-lobe of human serum transferrin. A structural study. J Biol Chem 278:6027–6033
3. Agosta F, Pievani M, Sala S, Geroldi C, Galluzzi S et al (2011) White matter damage in Alzheimer disease and its relationship to gray matter atrophy. Radiology 258:853–863
4. Ahmad KA, Ahmann JR, Migas MC, Waheed A, Britton RS et al (2002) Decreased liver hepcidin expression in the Hfe knockout mouse. Blood Cells Mol Dis 29:361–366
5. Aisen P, Leibman A (1972) Lactoferrin and transferrin: a comparative study. Biochim Biophys Acta 257:314–323
6. Ali-Rahmani F, Schengrund CL, Connor JR (2014) HFE gene variants, iron, and lipids: a novel connection in Alzheimer's disease. Front Pharmacol 5:165
7. Alimonti A, Ristori G, Giubilei F, Stazi MA, Pino A et al (2007) Serum chemical elements and oxidative status in Alzheimer's disease, Parkinson disease and multiple sclerosis. Neurotoxicology 28:450–456
8. Altamura S, Muckenthaler MU (2009) Iron toxicity in diseases of aging: Alzheimer's disease, Parkinson's disease and atherosclerosis. J Alzheimers Dis 16:879–895

9. Alzheimer's A (2016) 2016 Alzheimer's disease facts and figures. Alzheimers Dement 12:459–509
10. Anderson ED, VanSlyke JK, Thulin CD, Jean F, Thomas G (1997) Activation of the furin endoprotease is a multiple-step process: requirements for acidification and internal propeptide cleavage. EMBO J 16:1508–1518
11. Andrasi E, Farkas E, Gawlik D, Rosick U, Bratter P (2000) Brain iron and zinc contents of German patients with Alzheimer disease. J Alzheimers Dis 2:17–26
12. Arendt T (2009) Synaptic degeneration in Alzheimer's disease. Acta Neuropathol 118:167–179
13. Arnold RR, Cole MF, McGhee JR (1977) A bactericidal effect for human lactoferrin. Science 197:263–265
14. Arosio P, Ingrassia R, Cavadini P (2009) Ferritins: a family of molecules for iron storage, antioxidation and more. Biochim Biophys Acta 1790:589–599
15. Ayton S, Faux NG, Bush AI, Alzheimer's Disease Neuroimaging I. (2015) Ferritin levels in the cerebrospinal fluid predict Alzheimer's disease outcomes and are regulated by APOE. Nat Commun 6:6760
16. Ayton S, Fazlollahi A, Bourgeat P, Raniga P, Ng A et al (2017) Cerebral quantitative susceptibility mapping predicts amyloid-β-related cognitive decline. Brain 140:2112–2119
17. Baker EN, Baker HM (2009) A structural framework for understanding the multifunctional character of lactoferrin. Biochimie 91:3–10
18. Baker EN, Baker HM, Kidd RD (2002) Lactoferrin and transferrin: functional variations on a common structural framework. Biochem Cell Biol 80:27–34
19. Bandyopadhyay S, Hartley DM, Cahill CM, Lahiri DK, Chattopadhyay N, Rogers JT (2006) Interleukin-1alpha stimulates non-amyloidogenic pathway by alpha-secretase (ADAM-10 and ADAM-17) cleavage of APP in human astrocytic cells involving p38 MAP kinase. J Neurosci Res 84:106–118
20. Barnham KJ, Masters CL, Bush AI (2004) Neurodegenerative diseases and oxidative stress. Nat Rev Drug Discov 3:205–214
21. Bartzokis G (2011) Alzheimer's disease as homeostatic responses to age-related myelin breakdown. Neurobiol Aging 32:1341–1371
22. Bateman RJ, Aisen PS, De Strooper B, Fox NC, Lemere CA et al (2011) Autosomal-dominant Alzheimer's disease: a review and proposal for the prevention of Alzheimer's disease. Alzheimers Res Ther 3:1
23. Baum L, Chan IH, Cheung SK, Goggins WB, Mok V et al (2010) Serum zinc is decreased in Alzheimer's disease and serum arsenic correlates positively with cognitive ability. Biometals 23:173–179
24. Belaidi AA, Bush AI (2016) Iron neurochemistry in Alzheimer's disease and Parkinson's disease: targets for therapeutics. J Neurochem 139(Suppl 1):179–197
25. Berens C, Streicher B, Schroeder R, Hillen W (1998) Visualizing metal-ion-binding sites in group I introns by iron(II)-mediated Fenton reactions. Chem Biol 5:163–175
26. Bhatt L, Murphy C, O'Driscoll LS, Carmo-Fonseca M, McCaffrey MW, Fleming JV (2010) N-glycosylation is important for the correct intracellular localization of HFE and its ability to decrease cell surface transferrin binding. FEBS J 277:3219–3234
27. Bocca B, Forte G, Petrucci F, Pino A, Marchione F et al (2005) Monitoring of chemical elements and oxidative damage in patients affected by Alzheimer's disease. Ann Ist Super Sanita 41:197–203
28. Boddaert N, Le Quan Sang KH, Rotig A, Leroy-Willig A, Gallet S et al (2007) Selective iron chelation in Friedreich ataxia: biologic and clinical implications. Blood 110:401–408
29. Borgna-Pignatti C, Marsella M (2015) Iron chelation in Thalassemia major. Clin Ther 37:2866–2877
30. Bostrom F, Hansson O, Gerhardsson L, Lundh T, Minthon L et al (2009) CSF Mg and Ca as diagnostic markers for dementia with Lewy bodies. Neurobiol Aging 30:1265–1271
31. Bou-Abdallah F, Yang H, Awomolo A, Cooper B, Woodhall MR et al (2014) Functionality of the three-site ferroxidase center of Escherichia coli bacterial ferritin (EcFtnA). Biochemistry 53:483–495

32. Bousejra-ElGarah F, Bijani C, Coppel Y, Faller P, Hureau C (2011) Iron(II) binding to amyloid-β, the Alzheimer's peptide. Inorg Chem 50:9024–9030
33. Bradley-Whitman MA, Timmons MD, Beckett TL, Murphy MP, Lynn BC, Lovell MA (2014) Nucleic acid oxidation: an early feature of Alzheimer's disease. J Neurochem 128:294–304
34. Brickman AM, Provenzano FA, Muraskin J, Manly JJ, Blum S et al (2012) Regional white matter hyperintensity volume, not hippocampal atrophy, predicts incident Alzheimer disease in the community. Arch Neurol 69:1621–1627
35. Brickman AM, Zahodne LB, Guzman VA, Narkhede A, Meier IB et al (2015) Reconsidering harbingers of dementia: progression of parietal lobe white matter hyperintensities predicts Alzheimer's disease incidence. Neurobiol Aging 36:27–32
36. Brown JP, Hewick RM, Hellstrom I, Hellstrom KE, Doolittle RF, Dreyer WJ (1982) Human melanoma-associated antigen p97 is structurally and functionally related to transferrin. Nature 296:171–173
37. Burdo JR, Connor JR (2003) Brain iron uptake and homeostatic mechanisms: an overview. Biometals 16:63–75
38. Burns A, Zaudig M (2002) Mild cognitive impairment in older people. Lancet 360:1963–1965
39. Bush AI (2003) The metallobiology of Alzheimer's disease. Trends Neurosci 26:207–214
40. Butterfield DA, Drake J, Pocernich C, Castegna A (2001) Evidence of oxidative damage in Alzheimer's disease brain: central role for amyloid β-peptide. Trends Mol Med 7:548–554
41. Cai Z, Xiao M (2016) Oligodendrocytes and Alzheimer's disease. Int J Neurosci 126:97–104
42. Cardozo-Pelaez F, Song S, Parthasarathy A, Hazzi C, Naidu K, Sanchez-Ramos J (1999) Oxidative DNA damage in the aging mouse brain. Mov Disord 14:972–980
43. Carlson ES, Tkac I, Magid R, O'Connor MB, Andrews NC et al (2009) Iron is essential for neuron development and memory function in mouse hippocampus. J Nutr 139:672–679
44. Carro E, Bartolome F, Bermejo-Pareja F, Villarejo-Galende A, Molina JA et al (2017) Early diagnosis of mild cognitive impairment and Alzheimer's disease based on salivary lactoferrin. Alzheimers Dement (Amst) 8:131–138
45. Carvajal FJ, Mattison HA, Cerpa W (2016) Role of NMDA receptor-mediated glutamatergic signaling in chronic and acute neuropathologies. Neural Plast 2016:2701526
46. Chakraborty S, Kaur S, Guha S, Batra SK (2012) The multifaceted roles of neutrophil gelatinase associated lipocalin (NGAL) in inflammation and cancer. Biochim Biophys Acta 1826:129–169
47. Chan YK, Sung HK, Sweeney G (2015) Iron metabolism and regulation by neutrophil gelatinase-associated lipocalin in cardiomyopathy. Clin Sci (Lond) 129:851–862
48. Cheah JH, Kim SF, Hester LD, Clancy KW, Patterson SE 3rd et al (2006) NMDA receptor-nitric oxide transmission mediates neuronal iron homeostasis via the GTPase Dexras1. Neuron 51:431–440
49. Cheignon C, Tomas M, Bonnefont-Rousselot D, Faller P, Hureau C, Collin F (2018) Oxidative stress and the amyloid β peptide in Alzheimer's disease. Redox Biol 14:450–564
50. Cho HH, Cahill CM, Vanderburg CR, Scherzer CR, Wang B et al (2010) Selective translational control of the Alzheimer amyloid precursor protein transcript by iron regulatory protein-1. J Biol Chem 285:31217–31232
51. Choi J, Lee HW, Suk K (2011) Increased plasma levels of lipocalin 2 in mild cognitive impairment. J Neurol Sci 305:28–33
52. Chong MS, Sahadevan S (2005) Preclinical Alzheimer's disease: diagnosis and prediction of progression. Lancet Neurol 4:576–579
53. Connor JR, Menzies SL, St Martin SM, Mufson EJ (1990) Cellular distribution of transferrin, ferritin, and iron in normal and aged human brains. J Neurosci Res 27:595–611
54. Connor JR, Menzies SL, St Martin SM, Mufson EJ (1992) A histochemical study of iron, transferrin, and ferritin in Alzheimer's diseased brains. J Neurosci Res 31:75–83
55. Connor JR, Snyder BS, Arosio P, Loeffler DA, LeWitt P (1995) A quantitative analysis of isoferritins in select regions of aged, parkinsonian, and Alzheimer's diseased brains. J Neurochem 65:717–724

56. Connor JR, Snyder BS, Beard JL, Fine RE, Mufson EJ (1992) Regional distribution of iron and iron-regulatory proteins in the brain in aging and Alzheimer's disease. J Neurosci Res 31:327–335
57. Cooper LN, Bear MF (2012) The BCM theory of synapse modification at 30: interaction of theory with experiment. Nat Rev Neurosci 13:798–810
58. Cooper-Knock J, Kirby J, Ferraiuolo L, Heath PR, Rattray M, Shaw PJ (2012) Gene expression profiling in human neurodegenerative disease. Nat Rev Neurol 8:518–530
59. Cornett CR, Markesbery WR, Ehmann WD (1998) Imbalances of trace elements related to oxidative damage in Alzheimer's disease brain. Neurotoxicology 19:339–345
60. Corrigan FM, Reynolds GP, Ward NI (1993) Hippocampal tin, aluminum and zinc in Alzheimer's disease. Biometals 6:149–154
61. Corsi B, Cozzi A, Arosio P, Drysdale J, Santambrogio P et al (2002) Human mitochondrial ferritin expressed in HeLa cells incorporates iron and affects cellular iron metabolism. J Biol Chem 277:22430–22437
62. Crapper McLachlan DR, Dalton AJ, Kruck TP, Bell MY, Smith WL et al (1991) Intramuscular desferrioxamine in patients with Alzheimer's disease. Lancet 337:1304–1308
63. Crespo AC, Silva B, Marques L, Marcelino E, Maruta C et al (2014) Genetic and biochemical markers in patients with Alzheimer's disease support a concerted systemic iron homeostasis dysregulation. Neurobiol Aging 35:777–785
64. Cristovao JS, Santos R, Gomes CM (2016) Metals and neuronal metal binding proteins implicated in Alzheimer's disease. Oxid Med Cell Longev 2016:9812178
65. Dekens DW, Naude PJ, Engelborghs S, Vermeiren Y, Van Dam D et al (2017) Neutrophil gelatinase-associated lipocalin and its receptors in Alzheimer's disease (AD) brain regions: differential findings in AD with and without depression. J Alzheimers Dis 55:763–776
66. Devireddy LR, Gazin C, Zhu X, Green MR (2005) A cell-surface receptor for lipocalin 24p3 selectively mediates apoptosis and iron uptake. Cell 123:1293–1305
67. Dhennin-Duthille I, Masson M, Damiens E, Fillebeen C, Spik G, Mazurier J (2000) Lactoferrin upregulates the expression of CD4 antigen through the stimulation of the mitogen-activated protein kinase in the human lymphoblastic T Jurkat cell line. J Cell Biochem 79:583–593
68. Ding B, Chen KM, Ling HW, Sun F, Li X et al (2009) Correlation of iron in the hippocampus with MMSE in patients with Alzheimer's disease. J Magn Reson Imaging 29:793–798
69. Ding Q, Dimayuga E, Markesbery WR, Keller JN (2004) Proteasome inhibition increases DNA and RNA oxidation in astrocyte and neuron cultures. J Neurochem 91:1211–1218
70. Diouf I, Fazlollahi A, Bush AI, Ayton S, Alzheimer's disease Neuroimaging I. (2019) Cerebrospinal fluid ferritin levels predict brain hypometabolism in people with underlying β-amyloid pathology. Neurobiol Dis 124:335–339
71. Dixit R, Ross JL, Goldman YE, Holzbaur EL (2008) Differential regulation of dynein and kinesin motor proteins by tau. Science 319:1086–1089
72. Dixon SJ (2017) Ferroptosis: bug or feature? Immunol Rev 277:150–157
73. Dixon SJ, Lemberg KM, Lamprecht MR, Skouta R, Zaitsev EM et al (2012) Ferroptosis: an iron-dependent form of nonapoptotic cell death. Cell 149:1060–1072
74. De Domenico I, McVey Ward D, Kaplan J (2008) Regulation of iron acquisition and storage: consequences for iron-linked disorders. Nat Rev Mol Cell Biol 9:72–81
75. Duce JA, Tsatsanis A, Cater MA, James SA, Robb E et al (2010) Iron-export ferroxidase activity of β-amyloid precursor protein is inhibited by zinc in Alzheimer's disease. Cell 142:857–867
76. van Duijn S, Bulk M, van Duinen SG, Nabuurs RJA, van Buchem MA et al (2017) Cortical iron reflects severity of Alzheimer's disease. J Alzheimers Dis 60:1533–1545
77. Dunn LL, Sekyere EO, Suryo Rahmanto Y, Richardson DR (2006) The function of melanotransferrin: a role in melanoma cell proliferation and tumorigenesis. Carcinogenesis 27:2157–2169
78. Dusek P, Schneider SA, Aaseth J (2016) Iron chelation in the treatment of neurodegenerative diseases. J Trace Elem Med Biol 38:81–92

79. Elfawy HA, Das B (2019) Crosstalk between mitochondrial dysfunction, oxidative stress, and age related neurodegenerative disease: etiologies and therapeutic strategies. Life Sci 218:165–184
80. Esteves IM, Lopes-Aguiar C, Rossignoli MT, Ruggiero RN, Broggini ACS et al (2017) Chronic nicotine attenuates behavioral and synaptic plasticity impairments in a streptozotocin model of Alzheimer's disease. Neuroscience 353:87–97
81. Estrada JA, Contreras I, Pliego-Rivero FB, Otero GA (2014) Molecular mechanisms of cognitive impairment in iron deficiency: alterations in brain-derived neurotrophic factor and insulin-like growth factor expression and function in the central nervous system. Nutr Neurosci 17:193–206
82. Fava A, Pirritano D, Plastino M, Cristiano D, Puccio G et al (2013) The effect of lipoic acid therapy on cognitive functioning in patients with Alzheimer's disease. J Neurodegener Dis 2013:454253
83. Feder JN, Gnirke A, Thomas W, Tsuchihashi Z, Ruddy DA et al (1996) A novel MHC class I-like gene is mutated in patients with hereditary haemochromatosis. Nat Genet 13:399–408
84. Feder JN, Penny DM, Irrinki A, Lee VK, Lebron JA et al (1998) The hemochromatosis gene product complexes with the transferrin receptor and lowers its affinity for ligand binding. Proc Natl Acad Sci U S A 95:1472–1477
85. Fischer FU, Wolf D, Scheurich A, Fellgiebel A, Alzheimer's Disease Neuroimaging I. (2015) Altered whole-brain white matter networks in preclinical Alzheimer's disease. Neuroimage Clin 8:660–666
86. Fonte J, Miklossy J, Atwood C, Martins R (2001) The severity of cortical Alzheimer's type changes is positively correlated with increased amyloid-β Levels: Resolubilization of amyloid-β with transition metal ion chelators. J Alzheimers Dis 3:209–219
87. Food MR, Sekyere EO, Richardson DR (2002) The soluble form of the membrane-bound transferrin homologue, melanotransferrin, inefficiently donates iron to cells via nonspecific internalization and degradation of the protein. Eur J Biochem 269:4435–4445
88. Fricker M, Tolkovsky AM, Borutaite V, Coleman M, Brown GC (2018) Neuronal cell death. Physiol Rev 98:813–880
89. Friedmann Angeli JP, Schneider M, Proneth B, Tyurina YY, Tyurin VA et al (2014) Inactivation of the ferroptosis regulator Gpx4 triggers acute renal failure in mice. Nat Cell Biol 16:1180–1191
90. Frisoni GB, Boccardi M, Barkhof F, Blennow K, Cappa S et al (2017) Strategic roadmap for an early diagnosis of Alzheimer's disease based on biomarkers. Lancet Neurol 16:661–676
91. Fujioka M, Taoka T, Matsuo Y, Mishima K, Ogoshi K et al (2003) Magnetic resonance imaging shows delayed ischemic striatal neurodegeneration. Ann Neurol 54:732–747
92. Gaasch JA, Lockman PR, Geldenhuys WJ, Allen DD, Van der Schyf CJ (2007) Brain iron toxicity: differential responses of astrocytes, neurons, and endothelial cells. Neurochem Res 32:1196–1208
93. Gabbita SP, Lovell MA, Markesbery WR (1998) Increased nuclear DNA oxidation in the brain in Alzheimer's disease. J Neurochem 71:2034–2040
94. Galasko DR, Peskind E, Clark CM, Quinn JF, Ringman JM et al (2012) Antioxidants for Alzheimer disease: a randomized clinical trial with cerebrospinal fluid biomarker measures. Arch Neurol 69:836–841
95. Galazka-Friedman J, Bauminger ER, Szlachta K, Friedman A (2012) The role of iron in neurodegeneration–Mossbauer spectroscopy, electron microscopy, enzyme-linked immunosorbent assay and neuroimaging studies. J Phys: Condens Matter 24:244106
96. Gallagher JJ, Finnegan ME, Grehan B, Dobson J, Collingwood JF, Lynch MA (2012) Modest amyloid deposition is associated with iron dysregulation, microglial activation, and oxidative stress. J Alzheimers Dis 28:147–161
97. Gao M, Monian P, Pan Q, Zhang W, Xiang J, Jiang X (2016) Ferroptosis is an autophagic cell death process. Cell Res 26:1021–1032
98. Gerhardsson L, Blennow K, Lundh T, Londos E, Minthon L (2009) Concentrations of metals, β-amyloid and tau-markers in cerebrospinal fluid in patients with Alzheimer's disease. Dement Geriatr Cogn Disord 28:88–94

99. Gerstein M, Anderson BF, Norris GE, Baker EN, Lesk AM, Chothia C (1993) Domain closure in lactoferrin. Two hinges produce a see-saw motion between alternative close-packed interfaces. J Mol Biol 234:357–372
100. Giambattistelli F, Bucossi S, Salustri C, Panetta V, Mariani S et al (2012) Effects of hemochromatosis and transferrin gene mutations on iron dyshomeostasis, liver dysfunction and on the risk of Alzheimer's disease. Neurobiol Aging 33:1633–1641
101. Giovannelli L, Decorosi F, Dolara P, Pulvirenti L (2003) Vulnerability to DNA damage in the aging rat substantia nigra: a study with the comet assay. Brain Res 969:244–247
102. Gomes CM, Santos R (2013) Neurodegeneration in Friedreich's ataxia: from defective frataxin to oxidative stress. Oxid Med Cell Longev 2013:487534
103. Gomez-Pinilla F, Yang X (2018) System biology approach intersecting diet and cell metabolism with pathogenesis of brain disorders. Prog Neurobiol 169:76–90
104. Gomme PT, McCann KB, Bertolini J (2005) Transferrin: structure, function and potential therapeutic actions. Drug Discov Today 10:267–273
105. Gong NJ, Dibb R, Bulk M, van der Weerd L, Liu C (2019) Imaging β amyloid aggregation and iron accumulation in Alzheimer's disease using quantitative susceptibility mapping MRI. Neuroimage 191:176–185
106. Gonzalez J, Jurado-Coronel JC, Avila MF, Sabogal A, Capani F, Barreto GE (2015) NMDARs in neurological diseases: a potential therapeutic target. Int J Neurosci 125:315–327
107. Good PF, Perl DP, Bierer LM, Schmeidler J (1992) Selective accumulation of aluminum and iron in the neurofibrillary tangles of Alzheimer's disease: a laser microprobe (LAMMA) study. Ann Neurol 31:286–292
108. Goodman L (1953) Alzheimer's disease; a clinico-pathologic analysis of twenty-three cases with a theory on pathogenesis. J Nerv Ment Dis 118:97–130
109. Goozee K, Chatterjee P, James I, Shen K, Sohrabi HR et al (2018) Elevated plasma ferritin in elderly individuals with high neocortical amyloid-β load. Mol Psychiatry 23:1807–1812
110. Goralska M, Dackor R, Holley B, McGahan MC (2003) Alpha lipoic acid changes iron uptake and storage in lens epithelial cells. Exp Eye Res 76:241–248
111. Goswami T, Andrews NC (2006) Hereditary hemochromatosis protein, HFE, interaction with transferrin receptor 2 suggests a molecular mechanism for mammalian iron sensing. J Biol Chem 281:28494–28498
112. Griffiths PD, Crossman AR (1993) Distribution of iron in the basal ganglia and neocortex in postmortem tissue in Parkinson's disease and Alzheimer's disease. Dementia 4:61–65
113. Gu M, Owen AD, Toffa SE, Cooper JM, Dexter DT et al (1998) Mitochondrial function, GSH and iron in neurodegeneration and Lewy body diseases. J Neurol Sci 158:24–29
114. Guo C, Wang T, Zheng W, Shan ZY, Teng WP, Wang ZY (2013) Intranasal deferoxamine reverses iron-induced memory deficits and inhibits amyloidogenic APP processing in a transgenic mouse model of Alzheimer's disease. Neurobiol Aging 34:562–575
115. Guo C, Wang P, Zhong ML, Wang T, Huang XS et al (2013) Deferoxamine inhibits iron induced hippocampal tau phosphorylation in the Alzheimer transgenic mouse brain. Neurochem Int 62:165–172
116. Guo C, Yang ZH, Zhang S, Chai R, Xue H et al (2017) Intranasal lactoferrin enhances alpha-secretase-dependent amyloid precursor protein processing via the ERK1/2-CREB and HIF-1alpha pathways in an Alzheimer's disease mouse model. Neuropsychopharmacology 42:2504–2515
117. Guo C, Zhang YX, Wang T, Zhong ML, Yang ZH et al (2015) Intranasal deferoxamine attenuates synapse loss via up-regulating the P38/HIF-1alpha pathway on the brain of APP/PS1 transgenic mice. Front Aging Neurosci 7:104
118. Haeger P, Alvarez A, Leal N, Adasme T, Nunez MT, Hidalgo C (2010) Increased hippocampal expression of the divalent metal transporter 1 (DMT1) mRNA variants 1B and +IRE and DMT1 protein after NMDA-receptor stimulation or spatial memory training. Neurotox Res 17:238–247
119. Hallgren B, Sourander P (1958) The effect of age on the non-haemin iron in the human brain. J Neurochem 3:41–51

120. Hambright WS, Fonseca RS, Chen L, Na R, Ran Q (2017) Ablation of ferroptosis regulator glutathione peroxidase 4 in forebrain neurons promotes cognitive impairment and neurodegeneration. Redox Biol 12:8–17
121. Hamilton ML, Van Remmen H, Drake JA, Yang H, Guo ZM et al (2001) Does oxidative damage to DNA increase with age? Proc Natl Acad Sci U S A 98:10469–10474
122. Han M, Kim J (2015) Effect of dietary iron loading on recognition memory in growing rats. PLoS ONE 10:e0120609
123. Hare DJ, Doecke JD, Faux NG, Rembach A, Volitakis I et al (2015) Decreased plasma iron in Alzheimer's disease is due to transferrin desaturation. ACS Chem Neurosci 6:398–402
124. Heneka MT, Carson MJ, El Khoury J, Landreth GE, Brosseron F et al (2015) Neuroinflammation in Alzheimer's disease. Lancet Neurol 14:388–405
125. Hentze MW, Muckenthaler MU, Andrews NC (2004) Balancing acts: molecular control of mammalian iron metabolism. Cell 117:285–297
126. Heppner FL, Ransohoff RM, Becher B (2015) Immune attack: the role of inflammation in Alzheimer disease. Nat Rev Neurosci 16:358–372
127. Hirano T, Yamaguchi R, Asami S, Iwamoto N, Kasai H (1996) 8-hydroxyguanine levels in nuclear DNA and its repair activity in rat organs associated with age. J Gerontol A Biol Sci Med Sci 51:B303–B307
128. Hofer T, Perry G (2016) Nucleic acid oxidative damage in Alzheimer's disease-explained by the hepcidin-ferroportin neuronal iron overload hypothesis? J Trace Elem Med Biol 38:1–9
129. Holland R, McIntosh AL, Finucane OM, Mela V, Rubio-Araiz A et al (2018) Inflammatory microglia are glycolytic and iron retentive and typify the microglia in APP/PS1 mice. Brain Behav Immun 68:183–196
130. Honda K, Smith MA, Zhu X, Baus D, Merrick WC et al (2005) Ribosomal RNA in Alzheimer disease is oxidized by bound redox-active iron. J Biol Chem 280:20978–20986
131. Horowitz MP, Greenamyre JT (2010) Mitochondrial iron metabolism and its role in neurodegeneration. J Alzheimers Dis 20(Suppl 2):S551–S568
132. House E, Collingwood J, Khan A, Korchazkina O, Berthon G, Exley C (2004) Aluminium, iron, zinc and copper influence the in vitro formation of amyloid fibrils of $A\beta_{42}$ in a manner which may have consequences for metal chelation therapy in Alzheimer's disease. J Alzheimers Dis 6:291–301
133. House MJ, St Pierre TG, McLean C (2008) 1.4T study of proton magnetic relaxation rates, iron concentrations, and plaque burden in Alzheimer's disease and control postmortem brain tissue. Magn Reson Med 60:41–52
134. Hozumi I, Hasegawa T, Honda A, Ozawa K, Hayashi Y et al (2011) Patterns of levels of biological metals in CSF differ among neurodegenerative diseases. J Neurol Sci 303:95–99
135. Huang X, Atwood CS, Hartshorn MA, Multhaup G, Goldstein LE et al (1999) The Aβ peptide of Alzheimer's disease directly produces hydrogen peroxide through metal ion reduction. Biochemistry 38:7609–7616
136. Huang CW, Wang SJ, Wu SJ, Yang CC, Huang MW et al (2013) Potential blood biomarker for disease severity in the Taiwanese population with Alzheimer's disease. Am J Alzheimers Dis Other Demen 28:75–83
137. Hui Y, Wang D, Li W, Zhang L, Jin J et al (2011) Long-term overexpression of heme oxygenase 1 promotes tau aggregation in mouse brain by inducing tau phosphorylation. J Alzheimers Dis 26:299–313
138. Illing AC, Shawki A, Cunningham CL, Mackenzie B (2012) Substrate profile and metal-ion selectivity of human divalent metal-ion transporter-1. J Biol Chem 287:30485–30496
139. Ising C, Stanley M, Holtzman DM (2015) Current thinking on the mechanistic basis of Alzheimer's and implications for drug development. Clin Pharmacol Ther 98:469–471
140. Jiang T, Sun Q, Chen S (2016) Oxidative stress: A major pathogenesis and potential therapeutic target of antioxidative agents in Parkinson's disease and Alzheimer's disease. Prog Neurobiol 147:1–19
141. Juurlink BH, Thorburne SK, Hertz L (1998) Peroxide-scavenging deficit underlies oligodendrocyte susceptibility to oxidative stress. Glia 22:371–378

142. Kamalinia G, Khodagholi F, Atyabi F, Amini M, Shaerzadeh F et al (2013) Enhanced brain delivery of deferasirox-lactoferrin conjugates for iron chelation therapy in neurodegenerative disorders: in vitro and in vivo studies. Mol Pharm 10:4418–4431
143. Kamat PK, Kalani A, Rai S, Swarnkar S, Tota S et al (2016) Mechanism of oxidative stress and synapse dysfunction in the pathogenesis of Alzheimer's disease: understanding the therapeutics strategies. Mol Neurobiol 53:648–661
144. Kaur C, Ling EA (1999) Increased expression of transferrin receptors and iron in amoeboid microglial cells in postnatal rats following an exposure to hypoxia. Neurosci Lett 262:183–186
145. Kaur D, Sharma V, Deshmukh R (2019) Activation of microglia and astrocytes: a roadway to neuroinflammation and Alzheimer's disease. Inflammopharmacology
146. Kawabata H, Fleming RE, Gui D, Moon SY, Saitoh T et al (2005) Expression of hepcidin is down-regulated in TfR2 mutant mice manifesting a phenotype of hereditary hemochromatosis. Blood 105:376–381
147. Kawamata T, Tooyama I, Yamada T, Walker DG, McGeer PL (1993) Lactotransferrin immunocytochemistry in Alzheimer and normal human brain. Am J Pathol 142:1574–1585
148. Kennard ML, Feldman H, Yamada T, Jefferies WA (1996) Serum levels of the iron binding protein p97 are elevated in Alzheimer's disease. Nat Med 2:1230–1235
149. Kennard ML, Richardson DR, Gabathuler R, Ponka P, Jefferies WA (1995) A novel iron uptake mechanism mediated by GPI-anchored human p97. EMBO J 14:4178–4186
150. Kim AC, Lim S, Kim YK (2018) Metal ion effects on Aβ and Tau Aggregation. Int J Mol Sci 19
151. Krause A, Neitz S, Magert HJ, Schulz A, Forssmann WG et al (2000) LEAP-1, a novel highly disulfide-bonded human peptide, exhibits antimicrobial activity. FEBS Lett 480:147–150
152. Lane DJR, Ayton S, Bush AI (2018) Iron and Alzheimer's disease: an update on emerging mechanisms. J Alzheimers Dis 64:S379–S395
153. Langlois d'Estaintot B, Santambrogio P, Granier T, Gallois B, Chevalier JM et al (2004) Crystal structure and biochemical properties of the human mitochondrial ferritin and its mutant Ser144Ala. J Mol Biol 340:277–293
154. Lavados M, Guillon M, Mujica MC, Rojo LE, Fuentes P, Maccioni RB (2008) Mild cognitive impairment and Alzheimer patients display different levels of redox-active CSF iron. J Alzheimers Dis 13:225–232
155. Lavich IC, de Freitas BS, Kist LW, Falavigna L, Dargel VA et al (2015) Sulforaphane rescues memory dysfunction and synaptic and mitochondrial alterations induced by brain iron accumulation. Neuroscience 301:542–552
156. Lawrence CM, Ray S, Babyonyshev M, Galluser R, Borhani DW, Harrison SC (1999) Crystal structure of the ectodomain of human transferrin receptor. Science 286:779–782
157. Lee S, Viqar F, Zimmerman ME, Narkhede A, Tosto G et al (2016) White matter hyperintensities are a core feature of Alzheimer's disease: evidence from the dominantly inherited Alzheimer network. Ann Neurol 79:929–939
158. Lei P, Ayton S, Finkelstein DI, Spoerri L, Ciccotosto GD et al (2012) Tau deficiency induces parkinsonism with dementia by impairing APP-mediated iron export. Nat Med 18(2):291–295
159. Leveugle B, Spik G, Perl DP, Bouras C, Fillit HM, Hof PR (1994) The iron-binding protein lactotransferrin is present in pathologic lesions in a variety of neurodegenerative disorders: a comparative immunohistochemical analysis. Brain Res 650:20–31
160. Levi S, Finazzi D (2014) Neurodegeneration with brain iron accumulation: update on pathogenic mechanisms. Front Pharmacol 5:99
161. Levi S, Tiranti V (2019) Neurodegeneration with brain iron accumulation disorders: valuable models aimed at understanding the pathogenesis of iron deposition. Pharmaceuticals (Basel) 12
162. Li C, Lonn ME, Xu X, Maghzal GJ, Frazer DM et al (2012) Sustained expression of heme oxygenase-1 alters iron homeostasis in nonerythroid cells. Free Radic Biol Med 53:366–374
163. Lista S, O'Bryant SE, Blennow K, Dubois B, Hugon J et al (2015) Biomarkers in sporadic and familial Alzheimer's disease. J Alzheimers Dis 47:291–317

164. Liu JL, Fan YG, Yang ZS, Wang ZY, Guo C (2018) Iron and Alzheimer's disease: from pathogenesis to therapeutic implications. Front Neurosci 12:632
165. Liu Y, Lee SY, Neely E, Nandar W, Moyo M et al (2011) Mutant HFE H63D protein is associated with prolonged endoplasmic reticulum stress and increased neuronal vulnerability. J Biol Chem 286:13161–13170
166. Liu B, Moloney A, Meehan S, Morris K, Thomas SE et al (2011) Iron promotes the toxicity of amyloid β peptide by impeding its ordered aggregation. J Biol Chem 286:4248–4256
167. Livingston G, Sommerlad A, Orgeta V, Costafreda SG, Huntley J et al (2017) Dementia prevention, intervention, and care. Lancet 390:2673–2734
168. Loboda A, Jazwa A, Grochot-Przeczek A, Rutkowski AJ, Cisowski J et al (2008) Heme oxygenase-1 and the vascular bed: from molecular mechanisms to therapeutic opportunities. Antioxid Redox Signal 10:1767–1812
169. Lopes KO, Sparks DL, Streit WJ (2008) Microglial dystrophy in the aged and Alzheimer's disease brain is associated with ferritin immunoreactivity. Glia 56:1048–1060
170. Lovell MA, Gabbita SP, Markesbery WR (1999) Increased DNA oxidation and decreased levels of repair products in Alzheimer's disease ventricular CSF. J Neurochem 72:771–776
171. Lu T, Pan Y, Kao SY, Li C, Kohane I et al (2004) Gene regulation and DNA damage in the ageing human brain. Nature 429:883–891
172. Lu LN, Qian ZM, Wu KC, Yung WH, Ke Y (2017) Expression of iron transporters and pathological hallmarks of Parkinson's and Alzheimer's diseases in the brain of young, adult, and aged rats. Mol Neurobiol 54:5213–5224
173. Maamoun H, Benameur T, Pintus G, Munusamy S, Agouni A (2019) Crosstalk between oxidative stress and endoplasmic reticulum (ER) stress in endothelial dysfunction and aberrant angiogenesis associated with diabetes: a focus on the protective roles of heme oxygenase (HO)-1. Front Physiol 10:70
174. Magnani E, Fan J, Gasparini L, Golding M, Williams M et al (2007) Interaction of tau protein with the dynactin complex. EMBO J 26:4546–4554
175. Mairuae N, Hall Ii EC, Cheepsunthorn P, Lee SY, Connor JR (2010) The H63D HFE gene variant promotes activation of the intrinsic apoptotic pathway via mitochondria dysfunction following β-amyloid peptide exposure. J Neurosci Res 88:3079–3089
176. Mandel S, Amit T, Bar-Am O, Youdim MB (2007) Iron dysregulation in Alzheimer's disease: multimodal brain permeable iron chelating drugs, possessing neuroprotective-neurorescue and amyloid precursor protein-processing regulatory activities as therapeutic agents. Prog Neurobiol 82:348–360
177. Marro S, Chiabrando D, Messana E, Stolte J, Turco E et al (2010) Heme controls ferroportin1 (FPN1) transcription involving Bach1, Nrf2 and a MARE/ARE sequence motif at position—7007 of the FPN1 promoter. Haematologica 95:1261–1268
178. Mattson MP (2004) Metal-catalyzed disruption of membrane protein and lipid signaling in the pathogenesis of neurodegenerative disorders. Ann N Y Acad Sci 1012:37–50
179. Matute C (2010) Calcium dyshomeostasis in white matter pathology. Cell Calcium 47:150–157
180. McCarthy RC, Sosa JC, Gardeck AM, Baez AS, Lee CH, Wessling-Resnick M (2018) Inflammation-induced iron transport and metabolism by brain microglia. J Biol Chem 293:7853–7863
181. McKie AT, Barrow D, Latunde-Dada GO, Rolfs A, Sager G et al (2001) An iron-regulated ferric reductase associated with the absorption of dietary iron. Science 291:1755–1759
182. Meadowcroft MD, Connor JR, Smith MB, Yang QX (2009) MRI and histological analysis of β-amyloid plaques in both human Alzheimer's disease and APP/PS1 transgenic mice. J Magn Reson Imaging: JMRI 29:997–1007
183. Meadowcroft MD, Connor JR, Yang QX (2015) Cortical iron regulation and inflammatory response in Alzheimer's disease and APPSWE/PS1DeltaE9 mice: a histological perspective. Front Neurosci 9:255
184. Mecocci P, MacGarvey U, Beal MF (1994) Oxidative damage to mitochondrial DNA is increased in Alzheimer's disease. Ann Neurol 36:747–751

185. Mecocci P, Polidori MC, Ingegni T, Cherubini A, Chionne F et al (1998) Oxidative damage to DNA in lymphocytes from AD patients. Neurology 51:1014–1017
186. Mehlenbacher M, Poli M, Arosio P, Santambrogio P, Levi S et al (2017) Iron oxidation and core formation in recombinant heteropolymeric human ferritins. Biochemistry 56:3900–3912
187. Mena NP, Urrutia PJ, Lourido F, Carrasco CM, Nunez MT (2015) Mitochondrial iron homeostasis and its dysfunctions in neurodegenerative disorders. Mitochondrion 21:92–105
188. Di Meo I, Tiranti V (2018) Classification and molecular pathogenesis of NBIA syndromes. Eur J Paediatr Neurol 22:272–284
189. Mesquita SD, Ferreira AC, Falcao AM, Sousa JC, Oliveira TG et al (2014) Lipocalin 2 modulates the cellular response to amyloid β. Cell Death Differ 21:1588–1599
190. Migliore L, Fontana I, Trippi F, Colognato R, Coppede F et al (2005) Oxidative DNA damage in peripheral leukocytes of mild cognitive impairment and AD patients. Neurobiol Aging 26:567–573
191. Miyamoto T, Kashima H, Suzuki A, Kikuchi N, Konishi I et al (2011) Laser-captured microdissection-microarray analysis of the genes involved in endometrial carcinogenesis: stepwise up-regulation of lipocalin2 expression in normal and neoplastic endometria and its functional relevance. Hum Pathol 42:1265–1274
192. Molina JA, Jimenez-Jimenez FJ, Aguilar MV, Meseguer I, Mateos-Vega CJ et al (1998) Cerebrospinal fluid levels of transition metals in patients with Alzheimer's disease. J Neural Transm (Vienna) 105:479–488
193. Mouton-Liger F, Paquet C, Dumurgier J, Bouras C, Pradier L et al (2012) Oxidative stress increases BACE1 protein levels through activation of the PKR-eIF2alpha pathway. Biochim Biophys Acta 1822:885–896
194. Multhaup G, Schlicksupp A, Hesse L, Beher D, Ruppert T et al (1996) The amyloid precursor protein of Alzheimer's disease in the reduction of copper(II) to copper(I). Science 271:1406–1409
195. Munoz P, Humeres A (2012) Iron deficiency on neuronal function. Biometals 25:825–835
196. Munoz P, Humeres A, Elgueta C, Kirkwood A, Hidalgo C, Nunez MT (2011) Iron mediates N-methyl-D-aspartate receptor-dependent stimulation of calcium-induced pathways and hippocampal synaptic plasticity. J Biol Chem 286:13382–13392
197. Nakamichi N, Oikawa H, Kambe Y, Yoneda Y (2004) Relevant modulation by ferrous ions of N-methyl-D-aspartate receptors in ischemic brain injuries. Curr Neurovasc Res 1:429–440
198. Nandar W, Connor JR (2011) HFE gene variants affect iron in the brain. J Nutr 141:729S–739S
199. Nasrabady SE, Rizvi B, Goldman JE, Brickman AM (2018) White matter changes in Alzheimer's disease: a focus on myelin and oligodendrocytes. Acta Neuropathol Commun 6:22
200. Naude PJ, Nyakas C, Eiden LE, Ait-Ali D, van der Heide R et al (2012) Lipocalin 2: novel component of proinflammatory signaling in Alzheimer's disease. FASEB J 26:2811–2823
201. Nemeth E, Tuttle MS, Powelson J, Vaughn MB, Donovan A et al (2004) Hepcidin regulates cellular iron efflux by binding to ferroportin and inducing its internalization. Science 306:2090–2093
202. Nicolas G, Bennoun M, Devaux I, Beaumont C, Grandchamp B et al (2001) Lack of hepcidin gene expression and severe tissue iron overload in upstream stimulatory factor 2 (USF2) knockout mice. Proc Natl Acad Sci U S A 98:8780–8785
203. Nicolas G, Bennoun M, Porteu A, Mativet S, Beaumont C et al (2002) Severe iron deficiency anemia in transgenic mice expressing liver hepcidin. Proc Natl Acad Sci U S A 99:4596–4601
204. Nicolas G, Viatte L, Lou DQ, Bennoun M, Beaumont C et al (2003) Constitutive hepcidin expression prevents iron overload in a mouse model of hemochromatosis. Nat Genet 34:97–101
205. Niki E, Yoshida Y, Saito Y, Noguchi N (2005) Lipid peroxidation: mechanisms, inhibition, and biological effects. Biochem Biophys Res Commun 338:668–676
206. Nnah IC, Wessling-Resnick M (2018) Brain iron homeostasis: a focus on microglial iron. Pharmaceuticals (Basel) 11

207. Nunomura A, Chiba S, Lippa CF, Cras P, Kalaria RN et al (2004) Neuronal RNA oxidation is a prominent feature of familial Alzheimer's disease. Neurobiol Dis 17:108–113
208. Nunomura A, Perry G, Aliev G, Hirai K, Takeda A et al (2001) Oxidative damage is the earliest event in Alzheimer disease. J Neuropathol Exp Neurol 60:759–767
209. Nunomura A, Perry G, Pappolla MA, Friedland RP, Hirai K et al (2000) Neuronal oxidative stress precedes amyloid-β deposition in down syndrome. J Neuropathol Exp Neurol 59:1011–1017
210. Nunomura A, Perry G, Pappolla MA, Wade R, Hirai K et al (1999) RNA oxidation is a prominent feature of vulnerable neurons in Alzheimer's disease. J Neurosci 19:1959–1964
211. Ohgami RS, Campagna DR, Greer EL, Antiochos B, McDonald A et al (2005) Identification of a ferrireductase required for efficient transferrin-dependent iron uptake in erythroid cells. Nat Genet 37:1264–1269
212. Ou P, Tritschler HJ, Wolff SP (1995) Thioctic (lipoic) acid: a therapeutic metal-chelating antioxidant? Biochem Pharmacol 50:123–126
213. Ozcankaya R, Delibas N (2002) Malondialdehyde, superoxide dismutase, melatonin, iron, copper, and zinc blood concentrations in patients with Alzheimer disease: cross-sectional study. Croat Med J 43:28–32
214. Park CH, Valore EV, Waring AJ, Ganz T (2001) Hepcidin, a urinary antimicrobial peptide synthesized in the liver. J Biol Chem 276:7806–7810
215. Paul S, Connor JA (2010) NR2B-NMDA receptor-mediated increases in intracellular Ca^{2+} concentration regulate the tyrosine phosphatase, STEP, and ERK MAP kinase signaling. J Neurochem 114:1107–1118
216. Pelizzoni I, Macco R, Morini MF, Zacchetti D, Grohovaz F, Codazzi F (2011) Iron handling in hippocampal neurons: activity-dependent iron entry and mitochondria-mediated neurotoxicity. Aging Cell 10:172–183
217. Perez DR, Sklar LA, Chigaev A (2019) Clioquinol: to harm or heal. Pharmacol Ther
218. Persson T, Popescu BO, Cedazo-Minguez A (2014) Oxidative stress in Alzheimer's disease: why did antioxidant therapy fail? Oxid Med Cell Longev 2014:427318
219. Pierrel F, Cobine PA, Winge DR (2007) Metal Ion availability in mitochondria. Biometals 20:675–682
220. Pigeon C, Ilyin G, Courselaud B, Leroyer P, Turlin B et al (2001) A new mouse liver-specific gene, encoding a protein homologous to human antimicrobial peptide hepcidin, is overexpressed during iron overload. J Biol Chem 276:7811–7819
221. Pini L, Pievani M, Bocchetta M, Altomare D, Bosco P et al (2016) Brain atrophy in Alzheimer's disease and aging. Ageing Res Rev 30:25–48
222. Pooler AM, Noble W, Hanger DP (2014) A role for tau at the synapse in Alzheimer's disease pathogenesis. Neuropharmacology 76 Pt A:1–8
223. Pratico D, Uryu K, Leight S, Trojanoswki JQ, Lee VM (2001) Increased lipid peroxidation precedes amyloid plaque formation in an animal model of Alzheimer amyloidosis. J Neurosci 21:4183–4187
224. Puzzo D (2019) Abeta oligomers: role at the synapse. Aging (Albany NY) 11:1077–1078
225. Qian ZM, Shen X (2001) Brain iron transport and neurodegeneration. Trends Mol Med 7:103–108
226. Qian ZM, Wang Q (1998) Expression of iron transport proteins and excessive iron accumulation in the brain in neurodegenerative disorders. Brain Res Brain Res Rev 27:257–267
227. Radi E, Formichi P, Battisti C, Federico A (2014) Apoptosis and oxidative stress in neurodegenerative diseases. J Alzheimers Dis 42(Suppl 3):S125–S152
228. Radlowski EC, Johnson RW (2013) Perinatal iron deficiency and neurocognitive development. Front Hum Neurosci 7:585
229. Raha AA, Vaishnav RA, Friedland RP, Bomford A, Raha-Chowdhury R (2013) The systemic iron-regulatory proteins hepcidin and ferroportin are reduced in the brain in Alzheimer's disease. Acta Neuropathol Commun 1:55
230. Rajan MT, Jagannatha Rao KS, Mamatha BM, Rao RV, Shanmugavelu P et al (1997) Quantification of trace elements in normal human brain by inductively coupled plasma atomic emission spectrometry. J Neurol Sci 146:153–166

231. Rakba N, Aouad F, Henry C, Caris C, Morel I et al (1998) Iron mobilisation and cellular protection by a new synthetic chelator O-Trensox. Biochem Pharmacol 55:1797–1806
232. Ramos P, Santos A, Pinto NR, Mendes R, Magalhaes T, Almeida A (2014) Iron levels in the human brain: a post-mortem study of anatomical region differences and age-related changes. J Trace Elem Med Biol 28:13–17
233. Raven EP, Lu PH, Tishler TA, Heydari P, Bartzokis G (2013) Increased iron levels and decreased tissue integrity in hippocampus of Alzheimer's disease detected in vivo with magnetic resonance imaging. J Alzheimers Dis 37:127–136
234. Richardson DR (2000) The role of the membrane-bound tumour antigen, melanotransferrin (p97), in iron uptake by the human malignant melanoma cell. Eur J Biochem 267:1290–1298
235. Rishi G, Wallace DF, Subramaniam VN (2015) Hepcidin: regulation of the master iron regulator. Biosci Rep 35
236. Ritchie CW, Bush AI, Mackinnon A, Macfarlane S, Mastwyk M et al (2003) Metal-protein attenuation with iodochlorhydroxyquin (clioquinol) targeting Aβ amyloid deposition and toxicity in Alzheimer disease: a pilot phase 2 clinical trial. Arch Neurol 60:1685–1691
237. Robert A, Liu Y, Nguyen M, Meunier B (2015) Regulation of copper and iron homeostasis by metal chelators: a possible chemotherapy for Alzheimer's disease. Acc Chem Res 48:1332–1339
238. Rogers JT, Bush AI, Cho HH, Smith DH, Thomson AM et al (2008) Iron and the translation of the amyloid precursor protein (APP) and ferritin mRNAs: riboregulation against neural oxidative damage in Alzheimer's disease. Biochem Soc Trans 36:1282–1287
239. Roher AE, Weiss N, Kokjohn TA, Kuo YM, Kalback W et al (2002) Increased Aβ peptides and reduced cholesterol and myelin proteins characterize white matter degeneration in Alzheimer's disease. Biochemistry 41:11080–11090
240. Rottkamp CA, Raina AK, Zhu X, Gaier E, Bush AI et al (2001) Redox-active iron mediates amyloid-β toxicity. Free Radic Biol Med 30:447–450
241. Sabbir MG (2018) Loss of $Ca^{(2+)}$/calmodulin dependent protein kinase kinase 2 leads to aberrant transferrin phosphorylation and trafficking: a potential biomarker for Alzheimer's disease. Front Mol Biosci 5:99
242. Salvador GA (2010) Iron in neuronal function and dysfunction. BioFactors 36:103–110
243. Salvador GA, Uranga RM, Giusto NM (2010) Iron and mechanisms of neurotoxicity. Int J Alzheimers Dis 2011:720658
244. Santambrogio P, Biasiotto G, Sanvito F, Olivieri S, Arosio P, Levi S (2007) Mitochondrial ferritin expression in adult mouse tissues. J Histochem Cytochem 55:1129–1137
245. Sawikr Y, Yarla NS, Peluso I, Kamal MA, Aliev G, Bishayee A (2017) Neuroinflammation in Alzheimer's disease: the preventive and therapeutic potential of polyphenolic nutraceuticals. Adv Protein Chem Struct Biol 108:33–57
246. Scheltens P, Blennow K, Breteler MM, de Strooper B, Frisoni GB et al (2016) Alzheimer's disease. Lancet 388:505–517
247. Schipper HM, Song W (2015) A heme oxygenase-1 transducer model of degenerative and developmental brain disorders. Int J Mol Sci 16:5400–5419
248. Schipper HM, Song W, Tavitian A, Cressatti M (2019) The sinister face of heme oxygenase-1 in brain aging and disease. Prog Neurobiol 172:40–70
249. Schipper HM, Song W, Zukor H, Hascalovici JR, Zeligman D (2009) Heme oxygenase-1 and neurodegeneration: expanding frontiers of engagement. J Neurochem 110:469–485
250. Schneider C, Pratt DA, Porter NA, Brash AR (2007) Control of oxygenation in lipoxygenase and cyclooxygenase catalysis. Chem Biol 14:473–488
251. Schon EA, Area-Gomez E (2013) Mitochondria-associated ER membranes in Alzheimer disease. Mol Cell Neurosci 55:26–36
252. Schrag M, Mueller C, Oyoyo U, Smith MA, Kirsch WM (2011) Iron, zinc and copper in the Alzheimer's disease brain: a quantitative meta-analysis. Some insight on the influence of citation bias on scientific opinion. Prog Neurobiol 94:296–306
253. Seiler A, Schneider M, Forster H, Roth S, Wirth EK et al (2008) Glutathione peroxidase 4 senses and translates oxidative stress into 12/15-lipoxygenase dependent—and AIF-mediated cell death. Cell Metab 8:237–248

254. Sekyere EO, Dunn LL, Suryo Rahmanto Y, Richardson DR (2006) Role of melanotransferrin in iron metabolism: studies using targeted gene disruption in vivo. Blood 107:2599–2601
255. Shan X, Lin CL (2006) Quantification of oxidized RNAs in Alzheimer's disease. Neurobiol Aging 27:657–662
256. Shichiri M (2014) The role of lipid peroxidation in neurological disorders. J Clin Biochem Nutr 54:151–160
257. Shimada A, Keino H, Satoh M, Kishikawa M, Hosokawa M (2003) Age-related loss of synapses in the frontal cortex of SAMP10 mouse: a model of cerebral degeneration. Synapse 48:198–204
258. Shin RW, Kruck TP, Murayama H, Kitamoto T (2003) A novel trivalent cation chelator Feralex dissociates binding of aluminum and iron associated with hyperphosphorylated tau of Alzheimer's disease. Brain Res 961:139–146
259. da Silva VK, de Freitas BS, da Silva Dornelles A, Nery LR, Falavigna L et al (2014) Cannabidiol normalizes caspase 3, synaptophysin, and mitochondrial fission protein DNM1L expression levels in rats with brain iron overload: implications for neuroprotection. Mol Neurobiol 49:222–233
260. Smith MA, Harris PL, Sayre LM, Perry G (1997) Iron accumulation in Alzheimer disease is a source of redox-generated free radicals. Proc Natl Acad Sci U S A 94:9866–9868
261. Sperling RA, Aisen PS, Beckett LA, Bennett DA, Craft S et al (2011) Toward defining the preclinical stages of Alzheimer's disease: recommendations from the National Institute on Aging-Alzheimer's Association workgroups on diagnostic guidelines for Alzheimer's disease. Alzheimers Dement 7:280–292
262. Spires-Jones TL, Hyman BT (2014) The intersection of amyloid β and tau at synapses in Alzheimer's disease. Neuron 82:756–771
263. Squitti R, Ghidoni R, Scrascia F, Benussi L, Panetta V et al (2011) Free copper distinguishes mild cognitive impairment subjects from healthy elderly individuals. J Alzheimers Dis 23:239–248
264. Squitti R, Lupoi D, Pasqualetti P, Dal Forno G, Vernieri F et al (2002) Elevation of serum copper levels in Alzheimer's disease. Neurology 59:1153–1161
265. Squitti R, Salustri C, Siotto M, Ventriglia M, Vernieri F et al (2010) Ceruloplasmin/Transferrin ratio changes in Alzheimer's disease. Int J Alzheimers Dis 2011:231595
266. Srivastava RA, Jain JC (2002) Scavenger receptor class B type I expression and elemental analysis in cerebellum and parietal cortex regions of the Alzheimer's disease brain. J Neurol Sci 196:45–52
267. Stockwell BR, Friedmann Angeli JP, Bayir H, Bush AI, Conrad M et al (2017) Ferroptosis: a regulated cell death nexus linking metabolism, redox biology, and disease. Cell 171:273–285
268. Tang J, Oliveros A, Jang MH (2019) Dysfunctional mitochondrial bioenergetics and synaptic degeneration in Alzheimer disease. Int Neurourol J 23:S5–S10
269. Telling ND, Everett J, Collingwood JF, Dobson J, van der Laan G et al (2017) Iron biochemistry is correlated with amyloid plaque morphology in an established mouse model of Alzheimer's disease. Cell Chem Biol 24(1205–15):e3
270. Theil EC (2013) Ferritin: the protein nanocage and iron biomineral in health and in disease. Inorg Chem 52:12223–12233
271. Thomas G (2002) Furin at the cutting edge: from protein traffic to embryogenesis and disease. Nat Rev Mol Cell Biol 3:753–766
272. Thomsen MS, Andersen MV, Christoffersen PR, Jensen MD, Lichota J, Moos T (2015) Neurodegeneration with inflammation is accompanied by accumulation of iron and ferritin in microglia and neurons. Neurobiol Dis 81:108–118
273. Thorburne SK, Juurlink BH (1996) Low glutathione and high iron govern the susceptibility of oligodendroglial precursors to oxidative stress. J Neurochem 67:1014–1022
274. Tiwari S, Ndisang JF (2014) The heme oxygenase system and type-1 diabetes. Curr Pharm Des 20:1328–1337
275. Torsdottir G, Kristinsson J, Snaedal J, Johannesson T (2011) Ceruloplasmin and iron proteins in the serum of patients with Alzheimer's disease. Dement Geriatr Cogn Dis Extra 1:366–371

276. Tse KH, Herrup K (2017) DNA damage in the oligodendrocyte lineage and its role in brain aging. Mech Ageing Dev 161:37–50
277. Urrutia P, Aguirre P, Esparza A, Tapia V, Mena NP et al (2013) Inflammation alters the expression of DMT1, FPN1 and hepcidin, and it causes iron accumulation in central nervous system cells. J Neurochem 126:541–549
278. Urrutia PJ, Mena NP, Nunez MT (2014) The interplay between iron accumulation, mitochondrial dysfunction, and inflammation during the execution step of neurodegenerative disorders. Front Pharmacol 5:38
279. Ursini F, Bindoli A (1987) The role of selenium peroxidases in the protection against oxidative damage of membranes. Chem Phys Lipids 44:255–276
280. Valore EV, Ganz T (2008) Posttranslational processing of hepcidin in human hepatocytes is mediated by the prohormone convertase furin. Blood Cells Mol Dis 40:132–138
281. Varadarajan S, Yatin S, Aksenova M, Butterfield DA (2000) Review: Alzheimer's amyloid β-peptide-associated free radical oxidative stress and neurotoxicity. J Struct Biol 130:184–208
282. Veitch DP, Weiner MW, Aisen PS, Beckett LA, Cairns NJ et al (2019) Understanding disease progression and improving Alzheimer's disease clinical trials: recent highlights from the Alzheimer's disease neuroimaging initiative. Alzheimers Dement 15:106–152
283. Violet M, Chauderlier A, Delattre L, Tardivel M, Chouala MS et al (2015) Prefibrillar Tau oligomers alter the nucleic acid protective function of Tau in hippocampal neurons in vivo. Neurobiol Dis 82:540–551
284. Vural H, Demirin H, Kara Y, Eren I, Delibas N (2010) Alterations of plasma magnesium, copper, zinc, iron and selenium concentrations and some related erythrocyte antioxidant enzyme activities in patients with Alzheimer's disease. J Trace Elem Med Biol 24:169–173
285. Wally J, Halbrooks PJ, Vonrhein C, Rould MA, Everse SJ et al (2006) The crystal structure of iron-free human serum transferrin provides insight into inter-lobe communication and receptor binding. J Biol Chem 281:24934–24944
286. Walsh DM, Klyubin I, Fadeeva JV, Cullen WK, Anwyl R et al (2002) Naturally secreted oligomers of amyloid β protein potently inhibit hippocampal long-term potentiation in vivo. Nature 416:535–539
287. Wan L, Nie G, Zhang J, Zhao B (2012) Overexpression of human wild-type amyloid-β protein precursor decreases the iron content and increases the oxidative stress of neuroblastoma SH-SY5Y cells. J Alzheimers Dis 30:523–530
288. Wang J, Markesbery WR, Lovell MA (2006) Increased oxidative damage in nuclear and mitochondrial DNA in mild cognitive impairment. J Neurochem 96:825–832
289. Wang L, Sato H, Zhao S, Tooyama I (2010) Deposition of lactoferrin in fibrillar-type senile plaques in the brains of transgenic mouse models of Alzheimer's disease. Neurosci Lett 481:164–167
290. Wang J, Song N, Jiang H, Wang J, Xie J (2013) Pro-inflammatory cytokines modulate iron regulatory protein 1 expression and iron transportation through reactive oxygen/nitrogen species production in ventral mesencephalic neurons. Biochim Biophys Acta 1832:618–625
291. Wang X, Wang W, Li L, Perry G, Lee HG, Zhu X (2014) Oxidative stress and mitochondrial dysfunction in Alzheimer's disease. Biochim Biophys Acta 1842:1240–1247
292. Wang P, Wu Q, Wu W, Li H, Guo Y et al (2017) Mitochondrial ferritin deletion exacerbates β-amyloid-induced neurotoxicity in mice. Oxid Med Cell Longev 2017:1020357
293. Wang J, Xiong S, Xie C, Markesbery WR, Lovell MA (2005) Increased oxidative damage in nuclear and mitochondrial DNA in Alzheimer's disease. J Neurochem 93:953–962
294. Ward A, Hudson JW (2014) p53-Dependent and cell specific epigenetic regulation of the polo-like kinases under oxidative stress. PLoS ONE 9:e87918
295. Ward RJ, Zucca FA, Duyn JH, Crichton RR, Zecca L (2014) The role of iron in brain ageing and neurodegenerative disorders. Lancet Neurol 13:1045–1060
296. White RS, Bhattacharya AK, Chen Y, Byrd M, McMullen MF et al (2016) Lysosomal iron modulates NMDA receptor-mediated excitation via small GTPase, Dexras1. Mol Brain 9:38
297. Wimo A, Guerchet M, Ali GC, Wu YT, Prina AM et al (2017) The worldwide costs of dementia 2015 and comparisons with 2010. Alzheimers Dement 13:1–7

298. Wong BX, Duce JA (2014) The iron regulatory capability of the major protein participants in prevalent neurodegenerative disorders. Front Pharmacol 5:81
299. Wong BX, Tsatsanis A, Lim LQ, Adlard PA, Bush AI, Duce JA (2014) β-Amyloid precursor protein does not possess ferroxidase activity but does stabilize the cell surface ferrous iron exporter ferroportin. PLoS ONE 9:e114174
300. Wu WS, Zhao YS, Shi ZH, Chang SY, Nie GJ et al (2013) Mitochondrial ferritin attenuates β-amyloid-induced neurotoxicity: reduction in oxidative damage through the Erk/P38 mitogen-activated protein kinase pathways. Antioxid Redox Signal 18:158–169
301. Xu H, Jiang H, Xie J (2017) New insights into the crosstalk between NMDARs and iron: implications for understanding pathology of neurological diseases. Front Mol Neurosci 10:71
302. Xu H, Wang Y, Song N, Wang J, Jiang H, Xie J (2017) New progress on the role of glia in iron metabolism and iron-induced degeneration of dopamine neurons in Parkinson's disease. Front Mol Neurosci 10:455
303. Yamamoto A, Shin RW, Hasegawa K, Naiki H, Sato H et al (2002) Iron (III) induces aggregation of hyperphosphorylated tau and its reduction to iron (II) reverses the aggregation: implications in the formation of neurofibrillary tangles of Alzheimer's disease. J Neurochem 82:1137–1147
304. Yang H, Guan H, Yang M, Liu Z, Takeuchi S et al (2015) Upregulation of mitochondrial ferritin by proinflammatory cytokines: implications for a role in Alzheimer's disease. J Alzheimers Dis 45:797–811
305. Yang WS, SriRamaratnam R, Welsch ME, Shimada K, Skouta R et al (2014) Regulation of ferroptotic cancer cell death by GPX4. Cell 156:317–331
306. You LH, Yan CZ, Zheng BJ, Ci YZ, Chang SY et al (2017) Astrocyte hepcidin is a key factor in LPS-induced neuronal apoptosis. Cell Death Dis 8:e2676
307. Zatta P, Drago D, Bolognin S, Sensi SL (2009) Alzheimer's disease, metal ions and metal homeostatic therapy. Trends Pharmacol Sci 30:346–355
308. Zecca L, Stroppolo A, Gatti A, Tampellini D, Toscani M et al (2004) The role of iron and copper molecules in the neuronal vulnerability of locus coeruleus and substantia nigra during aging. Proc Natl Acad Sci U S A 101:9843–9848
309. Zhang YH, Wang DW, Xu SF, Zhang S, Fan YG et al (2018) alpha-Lipoic acid improves abnormal behavior by mitigation of oxidative stress, inflammation, ferroptosis, and tauopathy in P301S Tau transgenic mice. Redox Biol 14:535–548
310. Zilka N, Ferencik M, Hulin I (2006) Neuroinflammation in Alzheimer's disease: protector or promoter? Bratisl Lek Listy 107:374–383
311. Zmijewski JW, Landar A, Watanabe N, Dickinson DA, Noguchi N, Darley-Usmar VM (2005) Cell signalling by oxidized lipids and the role of reactive oxygen species in the endothelium. Biochem Soc Trans 33:1385–1389

Chapter 6
Iron Pathophysiology in Stroke

Mohammed M. A. Almutairi, Grace Xu and Honglian Shi

Abstract Ischemic and hemorrhagic stroke are the common types of stroke that lead to brain injury neurological deficits and mortality. All forms of stroke remain a serious health issue, and there is little successful development of drugs for treating stroke. Incomplete understanding of stroke pathophysiology is considered the main barrier that limits this research progress. Besides mitochondria and free radical-producing enzymes, labile iron is an important contributor to oxidative stress. Although iron regulation and metabolism in cerebral stroke are not fully understood, much progress has been achieved in recent years. For example, hepcidin has recently been recognized as the principal regulator of systemic iron homeostasis and a bridge between inflammation and iron regulation. This review discusses recent research progress in iron pathophysiology following cerebral stroke, focusing molecular regulation of iron metabolism and potential treatment targets.

Keywords Stroke · Iron · Hepcidin · FPN1 · HIF-1α · Anesthesia

6.1 Introduction

Stroke is one of the leading causes of death and long-term severe disability worldwide. Stroke can be classified into ischemic or hemorrhagic forms according to its pathophysiological nature. The majority of all stroke cases (around 90%) are ischemic, which occur due to a sudden decrease or loss of blood supply to the brain

M. M. A. Almutairi · H. Shi (✉)
Department of Pharmacology and Toxicology, School of Pharmacy, University of Kansas, Lawrence, KS 66045, USA
e-mail: hshi@ku.edu

M. M. A. Almutairi
Department of Pharmacology and Toxicology, School of Pharmacy, King Saud University, Riyadh 11451, Saudi Arabia

G. Xu
Department of Anesthesiology, School of Medicine, University of Kansas, Kansas City, KS 66160, USA

Y.-Z. Chang (ed.), *Brain Iron Metabolism and CNS Diseases*, Advances in Experimental Medicine and Biology 1173, https://doi.org/10.1007/978-981-13-9589-5_6

by thrombi and/or emboli. Intracerebral and/or subarachnoid rapturing blood vessels cause hemorrhagic stroke, comprising about 10% of all strokes [44, 82]. Even though the incidence percentage of hemorrhagic stroke is lower than ischemic stroke, it cannot be ignored because this type is more dangerous compared to the ischemic form [21]. In the last ten years, research has focused on identifying effective therapeutic approaches for reducing death and disability related to stroke. Numerous agents have been tested in animal models and clinical trials, but there has been limited progress in developing effective therapeutic strategies for stroke [14, 26, 104, 111]. An incomplete understanding of the mechanism responsible for stroke-induced neuronal injuries is a major issue underlying this limited progress. Disturbance of brain iron homeostasis contributes to neuronal injury following cerebral stroke [43, 49, 94]. Recently, there has been exciting progress in understanding iron metabolism in stroke. For example, hepcidin has recently been recognized as the central regulator of iron homeostasis and a bridge between inflammation and iron regulation [39, 83]. Many studies have shown that hepcidin enhances iron-dependent neuronal damage and subsequently has become a target for the development of novel therapeutics for iron toxicity in stroke [24, 101]. This review summarizes the pathophysiological role of iron in stroke as well as manipulation of iron-dependent pathways as a potential therapeutic approach to improve the outcome of stroke patients.

6.2 Brain Iron Regulation

Iron is essential for life as it is needed for many biological processes. In the brain, iron plays an important role in neurotransmission, myelination, cell division, and oxygen transport processes [110]. Brain cells maintain iron homeostasis by regulating iron import, storage, metabolism, and export [50, 51]. Many studies have demonstrated the role of transferrin, divalent metal transporter 1 (DMT1), ceruloplasmin (CP), ferritin, and ferroportin1 (FPN1) in the regulation of brain iron [70, 87]. Under physiological conditions, iron binds to transport proteins such as transferrin. Transferrin-bound iron is transported into brain cells via receptor-mediated endocytosis. Emerging evidence suggests that uptake of transferrin-bound iron into the brain cells by transferrin receptor-mediated endocytosis is the major known pathway of brain iron transport across the blood–brain barrier (BBB) [87]. This route of iron transport consists of several steps, including transferrin-iron binding, endocytosis of transferrin-bound iron, dissociation of iron from transferrin, and translocation of the iron inside brain cells. Unbound iron can gain access into brain cells through DMT1, but the majority of iron in brain cells is bound to storage proteins, such as ferritin. CP promotes iron release from intracellular stores. FPN1 is the first and solely known vertebrate iron exporter expressed in the brain. It has been shown that FPN1 mediates iron efflux from brain cells. Since iron import, storage, and export are tightly regulated, any alterations in these steps may disturb iron homeostasis in the brain resulting in iron neurotoxicity [2, 87]. Deficiency in FPN1 activity results in iron retention and further iron toxicity [28]. Increased free iron levels have been observed in ischemic and hemorrhagic brains; however, the exact mechanism behind it is not fully understood [16, 94].

6.3 Alterations of Iron Homeostasis in Ischemic and Hemorrhagic Stroke

Iron plays a critical role in oxygen transport [110]. Under ischemic conditions, required levels of oxygen in the brain increase, resulting in high demand on iron transport and metabolism inside preserved brain regions [94]. Ferritin-bound iron is released in response to changes from ischemic insults, including acidosis and inflammatory mediators. Many studies have shown that the binding affinity of transferrin for iron decreases at low pH levels, resulting in dissociation of iron from transferrin [64]. Upon dissociation, unbound iron can easily gain access to the extracellular space where it is transported into neurons, resulting in increased intracellular iron levels [79, 94]. Recent observations have demonstrated that decreased iron efflux from neuronal cells contributes to putative alterations in iron homeostasis following ischemic stroke [62]. Furthermore, growing evidence indicates that iron accumulation in neurons following ischemia increases brain susceptibility to iron-induced damage [41, 64, 94]. In hemorrhagic stroke condition, many studies showed iron overload in the brain after intracerebral hemorrhage (ICH) [16, 88]. Red blood cell release and lysis have been considered the major factors mediating iron-induced brain damage after ICH [53]. Collagenase, which is commonly used to induce ICH, results in iron overload [113]. Moreover, iron overload may contribute to brain edema following ICH [43]. However, the potential mechanism, underlying iron-mediated neurotoxicity, is unsatisfactorily understood.

6.4 Iron Toxicity

Many studies have suggested that the severity of neuronal damage after ischemic stroke is proportional to the magnitude of brain iron accumulation [94]. For example, Patt et al. demonstrated that gerbils with low brain iron exhibited fewer neurological deficits compared to those with high brain iron levels [81]. Consistently, rats with iron overload exhibited about 60% greater infarct volume than rats with lower iron load [12]. Iron accumulation in rat brains following ischemia [58, 94] induces lipid peroxidation and neuronal injury by catalyzing conversion of superoxide and hydrogen peroxide into highly reactive toxic hydroxyl radicals [54]. Increased formation of these free radicals results in oxidative stress and subsequent ischemic neuronal death [11]. In addition, it has been shown that experimental ICH causes brain injury via iron-induced oxidation and DNA damage [71, 114]. Results from animal studies indicated that Tirilazad mesylate, an inhibitor of iron-dependent lipid peroxidation, reduces neuronal injury in global and focal cerebral ischemia [31, 47]. Moreover, deferoxamine (DFO), a potent iron-chelating agent, reduces infract volume and neurological deficits in ischemic rats [48, 97] and human patients [93]. Rosenthal et al. have reported that DFO improves neurological status and reduces mortality rate in rats subjected to a 6.5 min-cardiac arrest [90]. DFO also improves neurologic function in collagenase-induced ICH model [113]. Together, these results support the notion that iron induces neurotoxicity following cerebral stroke.

6.5 Molecular Mechanisms Involved in Iron Toxicity

6.5.1 Ferroportin in Ischemic and ICH Stroke

Ferroportin 1 (FPN1, encoded by SLC40A1) is a transmembrane protein, which contains 12 domains and is expressed in the duodenal enterocytes, placenta, macrophages, and brain [1, 7, 67]. In the brain, FPN1 is predominately expressed in neurons of the cerebral cortex, hippocampus, and cerebellum [7, 9] where it functions as the sole protein capable of exporting iron from cells [37]. In 2000, Donovan et al. found that a complete loss of FPN1 resulted in severe iron deficiency anemia and caused embryonic death in zebrafish and mammals [27]. Additionally, FPN1-mutated mice exhibited severe cellular iron accumulation [28, 125]. Emerging evidence shows that ischemic stroke disrupted the mechanisms that maintaining brain iron homeostasis and further resulted in iron neurotoxicity [56]. Li et al. demonstrated that FPN1 expression was down-regulated in the cerebral cortex and hippocampus after cerebral ischemia. A decrease in FPN1 expression in an ischemic brain may result in iron overload and hepcidin synthesis [62]. Since brain endothelial cells express FPN1, this transporter plays an important role in ICH-induced iron overload. Many studies have showed that FPN1 significantly increased after ICH [107]. Despite the largely unknown neurophysiological role of FPN1 in the brain, FPN1 has been shown to act as a receptor for hepcidin in cultured neuronal cells [74].

6.5.2 Hepcidin

Hepcidin was first discovered in human blood and urine as a bactericidal peptide synthesized in the liver [46, 55]. Since its discovery, many reports have demonstrated that hepcidin plays a critical role in systemic regulation of iron metabolism in mammals including human, rat, mouse, and pig [33, 50, 72]. The gene encoding hepcidin (HAMP) is expressed in a growing list of tissues, including liver, brain, lungs, intestines, and heart [15, 35, 38, 68, 109]. Hepcidin is a central regulator of iron homeostasis, but the exact mechanism of how hepcidin regulates iron acquisition has not been fully characterized. Nicolas et al. unambiguously recognized that inactivation of HAMP results in severe iron accumulation in the liver and pancreas [76]. Additionally, transgenic mice overexpressing HAMP exhibit a severe anemia [77]. Human patients with hepcidin mutations suffer from juvenile hemochromatosis [89]. Treating cultured cells expressing an FPN1-GFP with hepcidin results in translocation of FPN1-GFP from the cell surface into lysosomes where it is degraded [19]. Moreover, the injection of hepcidin into the cerebral ventricle suppresses FPN1 protein levels in different regions of the brain, such as the cerebral cortex, hippocampus, and striatum [24]. Xiong et al. reported that both brain and serum hepcidin was up-regulated after ICH. This increased expression of hepcidin reduces intracellular iron efflux from brain microvessel endothelial cells via regulating FPN1 [117]. Studies have shown that hepcidin binds to FPN1 and induces intracellular iron accumulation via interaction with FPN1 and causes its internalization and degradation in endolysosomes (Fig. 6.1) [19, 20]. It has been identified that residue Cys326 of FPN1 is required for hepcidin interaction and mutation of this residue induces FPN1 resistance to hepcidin binding [20]. Fernandes et al. proposed that a thiol subunit of the Cys326 residue initiates a disulfide bond with a cysteine residue in hepcidin [17, 32]. These findings indicate that increased iron accumulation and decreased iron efflux might be caused by the action of hepcidin. In physiological conditions, intercellular iron is mainly stored in ferritin and FPN1 mediates efflux of un-needed or excess iron. During many pathological conditions, including stroke, hepcidin is

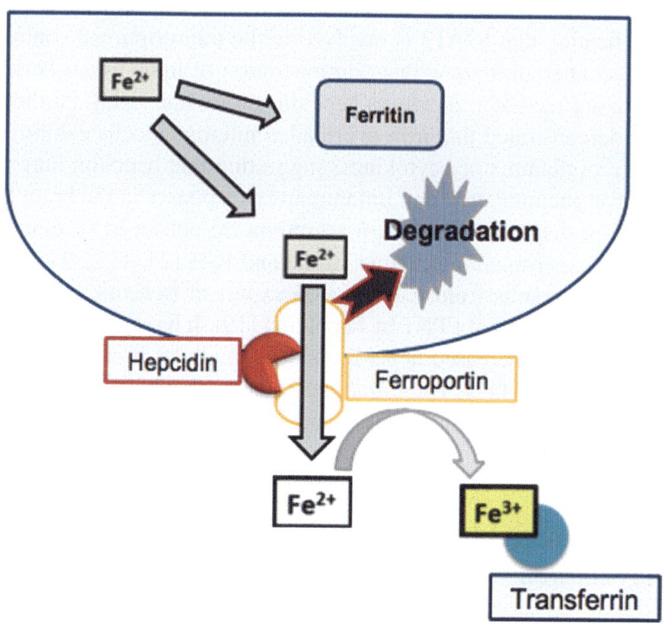

Fig. 6.1 Mechanism of hepcidin action. Hepcidin increases intracellular iron levels by causing the degradation of FPN1, a cellular membrane protein that export iron to extracellular space

up-regulated and binds to FPN1 and causes its internalization and degradation in endolysosomes. FPN1 degradation leads to iron accumulation in neurons and further results in neuronal death.

6.5.3 *Regulation of Hepcidin–FPN1 Pathway*

Hepcidin expression is regulated and modulated at the transcriptional level by many factors, including iron status and inflammation [61]. For example, the inflammatory cytokine interleukin (IL)-6, a major inflammatory mediator, up-regulates hepcidin expression and reduces serum iron levels in humans and experimental animal models [73, 112]. Additional studies have suggested that hepcidin expression is up-regulated by inflammatory cytokines following infection and post-ischemic inflammatory conditions [1, 105]. The mechanisms underlying inflammation-induced hepcidin expression are complex and only partially understood. There is increasing evidence that inflammation positively regulates hepcidin expression, mainly through induction of signal transducer and activator of transcription3 (STAT3) pathway [106]. Upon inflammatory stimuli, IL-6 activates Janus Kinase (JAK), a kinase responsible for the STAT3 phosphorylation. Phosphorylated STAT3 (p-STAT3) translocates into the nucleus and binds its response element in the hepcidin promoter, resulting in the stimulation of hepcidin transcription (Fig. 6.2) [112]. At transcriptional levels, knockdown of STAT3 significantly down-regulates endogenous hepcidin mRNA expression, indicating that STAT3 is involved in the transcriptional control of hepcidin [106]. Recent studies show that administration of lipopolysaccharide (LPS), a key inflammatory mediator, increases hepcidin expression [108]. Furthermore, an in-vitro study demonstrated that iron-overloaded microglial cells exhibit increased production of proinflammatory cytokines, suggesting that hepcidin may provide a link between iron accumulation and inflammatory responses in the brain [92, 105]. Furthermore, iron disposition and IL-6 secretion are enhanced in mouse and rat brains following experimental ischemic stroke and ICH [24, 115]. This intracellular iron disposition results from reduced iron export in ischemic brain due to the down-regulated expression of FPN1 by hepcidin [119]. It has also been reported that ICH models exhibited increased hepcidin and intercellular iron accumulation due to decreased iron efflux [117]. These findings indicate that hepcidin may promote a crucial association between inflammation and iron overload in brain tissues.

6.5.4 *Iron-Mediated Hypoxia-Inducible Factor 1α (HIF-1α) Degradation*

HIF-1 has been suggested to be a critical regulator of neurological outcomes following ischemic stroke due to its function as a central role in regulating neuronal adap-

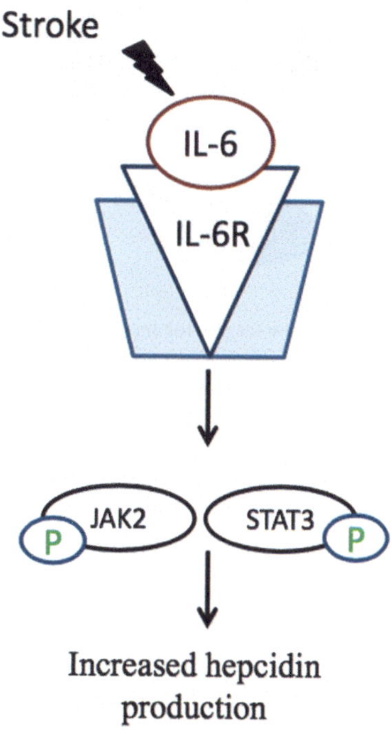

Fig. 6.2 Inflammation-induced hepcidin transcription. Inflammation plays a key role in upregulating hepcidin expression. Stroke initiates the production of inflammatory stimuli. Upon inflammatory stimuli, interleukin (IL)-6 activates Janus Kinase (JAK), a kinase responsible for the STAT3 phosphorylation. Phosphorylated STAT3 (p-STAT3) translocates into the nucleus and binds its response element in the hepcidin promoter, resulting in the stimulation of hepcidin transcription action

tive responses to hypoxia. HIF-1 regulates genes that promote energy metabolism, angiogenesis, and erythropoiesis, all of which are critical in promoting cell survival in hypoxia. HIF-1 is a heterodimer, consisting of a constitutively expressed β subunit and a regulated α subunit. HIF-1α is preferentially degraded under normoxic conditions while under hypoxic conditions, it is stabilized, dimerizes with the β subunit, and subsequently induces adaptive gene expression and promotes cell survival. Under normoxic conditions, HIF-1α is degraded through the ubiquitin–proteasome system after recognized by the von Hippel Lindau tumor suppressor (pVHL), a component of the E3 ubiquitin ligase. The recognition by pVHL requires the action of a family of HIF prolyl hydroxylase domain (PHD) that hydroxylates proline residues on HIF-1α in the presence of oxygen. Ferrous ion is a cofactor of the enzyme PHD. Increase in free iron promotes HIF-1α degradation through the 26S pathway by elevating its hydroxylation and thus compromises cellular resistance to ischemic injury (Fig. 6.3). Conversely, reducing iron levels would be lower PHD activity and increase HIF-1α stability/accumulation. In fact, Hishikawa et al. have reported that DFO significantly attenuated basilar artery vasospasm and reduction of brainstem blood flow in a rat model of hemorrhagic stroke by increase in HIF-1α protein level and HIF-1 activity [52]. Similarly, in ischemic stroke DFO induced HIF-1 DNA binding and transcription of erythropoietin in vivo and ischemic tolerance [5, 86]. More recently, Sorond et al. reported that DFO infusion improved cerebrovascular function in older individuals, possibly by increased the activity of HIF-1-regulated pathways [99]. Although it is arguable if the protective effect of DFO can be all attributed to

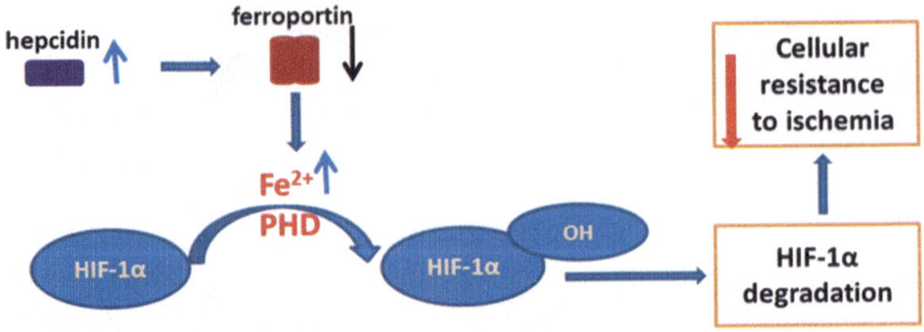

Fig. 6.3 New targets for increasing HIF-1 activity. Iron reduction could be an effective pharmacological approach to activate HIF-1 and to enhance cerebrovascular function acute and chronic ischemic conditions. Hepcidin and FPN1 may provide additional targets for increasing HIF-1 levels

HIF-1 activity [122], iron reduction could be an effective pharmacological approach to activate HIF-1 and to enhance cerebrovascular function acute and chronic ischemic conditions. Hepcidin and FPN1 may provide additional targets for increasing HIF-1 levels (Fig. 6.3).

6.5.5 Post-ischemic and ICH Inflammation Regulates Neuronal Damage

Neuro-inflammation following ischemia plays a crucial role in neuronal injury [25]. The inflammatory response may promote cerebral swelling, which is often lethal [96]. Microglial cells, which are the resident macrophages of the brain, modulate neuronal damage by releasing several proinflammatory mediators such as cytokines and chemokines in response to post-ischemic inflammation [29, 59]. It has been reported that the number of activated microglia significantly increases in rat brains subjected to 2-h ischemia [65]. Many cytokines, including ILs and tumor necrosis factor-α (TNF-α), are up-regulated in ischemic brains [59, 118, 123]. Additionally, elevated cytokine levels following ischemia exaggerate the infarct severity and induce further neuronal death [59]. Intraventricular injection of recombinant IL-1β worsens the infract severity in ischemic mouse brains when compared with controls [118]. However, IL-1β deficient mice exhibit less neuronal damage in comparison with wild-type [8]. In addition to cytokines, chemokine levels are elevated in ischemic animal models while their deficiency rescues neurons from ischemic injury [57, 98]. Consistently, mice lacking chemokine receptors exhibit less brain injury after ischemia [23]. Similar to ischemic stroke, microglial activation, cytokines, and chemokines release have been reported in ICH [124]. These findings suggest that inflammation may contribute to neuronal injury and expand the brain lesion upon ischemia and ICH. Recent studies have reported that inflammation mediates iron accumulation in CNS cells [105]. As mentioned previously, increased iron deposition has been linked to neuronal damage following ischemia and ICH [94, 117]. These findings indicate that inflammation may drive iron accumulation in cerebral stroke. However, the hypothesis that hepcidin acts as a link between inflammation and iron neurotoxicity requires further investigation.

6.5.6 Effect of Anesthesia on Iron Regulation in Stroke

Postoperative cognitive dysfunction (POCD) refers to a decline in cognition after major surgery and anesthesia. Age is one major factor for cognitive complications following surgery [116]. Based on a prospective longitudinal study, Monk et al. found that stroke is another major independent factor that increases the incidence of prolonged POCD [69]. The authors argued that the increased incidence of prolonged POCD may be related to the concept of cognitive reserve because the patients may have lost critical neural mass with their stroke, leading to a decrease in their cognitive reserve and an increased susceptibility to POCD. Alternatively, the latest clinical evidence establishes a positive relationship between general anesthesia and development of POCD in the elderly [6, 22, 66, 95, 100]. It has been suggested that anesthetics may promote inflammation, increase oxidative stress, reduce neurotrophin expression, and cholinergic dysfunction, which contribute to POCD. We speculate that iron accumulation might be one contributing factor to anesthesia-mediated POCD due to the following factors. (1) Iron overload has direct links to key components of POCD symptoms such as in Alzheimer's diseases [3, 45]. (2) It seems that oxidative stress contributes to the progression of POCD, and iron overload can cause oxidative stress. (3) It has been reported that anesthetics could increase the production of inflammation mediators, which can upregulate hepcidin expression, causing iron overload. (4) Preconditioning is to stimulate the endogenous protection mechanism with a mild stress. Sevoflurane and isoflurane preconditioning ameliorates inflammation, cerebral lipid peroxidation, and histologic injury in a rat model of focal cerebral ischemia [4], implying that the prolonged anesthetic exposure could cause adverse effects such as inflammation and lipid peroxidation. Future studies are needed to explore the potential role of iron metabolism in anesthetics-mediated brain dysfunction such as POCD.

6.6 Iron as a Treatment Target

In the last ten years, our understanding of the pathogenesis of stroke-induced brain injury has advanced. However, therapeutic options for cerebral stroke are still very limited. Current approved therapies rely mainly on the use of thrombolytic treatment in acute ischemic cases [104]. Recombinant tissue plasminogen activator (r-tPA) is the only thrombolytic drug that has been approved for clinical use to reduce neurological injury and improve the functional recovery of patients with cerebral ischemia [91]. The time window for effective treatment with r-tPA is limited to three hours after symptom onset. Because the large majority of patients with ischemic stroke are not aware of initial symptoms, many do not seek care within the three-hour time frame. Consequently, due to its short therapeutic window and potential hemorrhagic risk, the number of patients that benefit from this treatment is limited [104]. In term of ICH, several reports have shown that majority of ICH-patients who received anticoagulant and antiplatelet treatments had experienced the incidence of severe anticoagulant-associated ICH [34]. In addition, emerging evidence demonstrated that antithrombotic drugs such as warfarin increased the risk of ICH by seven folds and is associated disability and death [30, 40]. This major challenge increases the demand for safe and effective therapeutic approaches to improve the outcome of this pathological condition. This section discusses some prospective therapeutic strategies that can protect the brain from the negative consequences of cerebral stroke. The negative consequences of stroke that could be beneficial therapeutic targets, including post-stroke inflammation and brain iron accumulation [94, 123, 124].

6.6.1 Anti-inflammatory Treatment Approach

As mentioned previously, post-ischemic inflammation is a large effector in the pathology of cerebral stroke [96, 123, 124]. Therefore, targeting inflammatory pathways can be a protective therapeutic strategy to reduce or limit brain injury after the stroke [104]. Many anti-inflammatory molecules have been tested as neuroprotective agents in acute ischemic stroke. Although most of these agents have failed to restore neurological functions in patients with ischemic stroke, blocking post-ischemic inflammation may improve pro-regenerative effects following ischemia. IL-10 and transforming growth factor beta (TGF-β) are major anti-inflammatory mediators in ischemic and ICH brain injury [42, 96, 103]. Many studies have been conducted to determine the beneficial effect of these anti-inflammatory mediators in ischemic stroke treatment. Ooboshi et al. reported that overexpression of IL-10 protects neurons against focal and global brain ischemia. Their results also found that IL-10-treated ischemic brains exhibit less brain infarction and immune cell infiltrations in comparison with controls [78]. Recently, Cekanaviciute et al. demonstrated that TGF-β inhibits subacute neuroinflammation and preserves brain function in the peri-infarct cortex after ischemic stroke [13]. Clinically, minocycline is one of the most anti-inflammatory drugs that have been tested in the treatment of human acute ischemic stroke and ICH [10, 104]. It has been found that minocycline improves the prognosis of patients or animals with stroke compared with placebo treatment [18, 60]. Although the exact mechanism as well as the onset effect of these anti-inflammatory molecules in ischemia treatment remain to be clarified, controlling post-ischemic inflammatory pathway shows a promising potential target to treat ischemic stroke [96]. The next section discusses whether iron-chelating agents could be a neuroprotective tool against inflammation-induced iron neurotoxicity in ischemic stroke.

6.6.2 Iron-Chelating Therapy

Considerable evidence suggests that iron-mediated neurotoxicity plays a critical role in the pathogenesis of ischemic and ICH stroke [43, 94]. Additionally, it has been found that iron accumulation in the brain is induced by post-ischemic and ICH inflammation [18, 24]. Based on these notions, many studies have examined whether decreasing iron overload in the brain during ischemia decreases neurological defects and improves the outcome of ischemic stroke. Iron chelators bind excess iron in a stable complex to prevent iron accumulation and maintain a safe iron status [84]. In fact, DFO reduces ischemic and ICH stroke damage in rodent models [75, 80]. Additionally, DFO significantly decreases cortical infarct volume and improves neurologic deficits following ischemia in rat brains [48]. Furthermore, these neuroprotective effects of DFO are mediated by suppression of iron-induced free radical formation in animal models after ischemic stroke [93]. As discussed previously in this review, iron accumulation, FPN1 reduction and hepcidin up-regulation are all induced during ischemia-induced neuroinflammation [24, 105]. DFO dramatically blocks increased iron levels and reduces FPN1 down-regulation in LPS-treated neurons[121]. These findings suggest that DFO may also attenuate neuroinflammation-induced iron accumulation in neurons upon stroke via regulation of hepcidin pathway.

6.6.3 Targeting Hepcidin Pathway in Ischemic and ICH Stroke

Hepcidin, a negative regulator of the sole cellular iron exporter FPN1, plays a crucial role in intracellular iron accumulation [19, 20]. Therefore, therapeutic approaches that decrease hepcidin synthesis and/or increase FPN1 activity seem to be potential candidates to treat patient with iron overload disorders [85]. Recent studies have showed that hepcidin siRNA treatment induces the FPN1 expression and rescues iron accumulation in cerebral cortex, hippocampus, and corpus striatum of ischemic mouse brains [24]. In addition, hepcidin knockout mice ICH models exhibit reduced brain iron accumulation compared with wild-type mice [117]. These findings suggest that iron overload following ischemia and ICH is negatively correlated with increased FPN1 levels in hepcidin knockout models. Additionally, alteration of hepcidin-FPN1 interaction with a thiol-reactive compound fursultiamine prevents FPN1 internalization and degradation via sequestering the Cys326 residue of FPN1, a necessary residue for hepcidin binding to FPN1 [85]. Consequently, fursultiamine reduces intracellular iron accumulation via enhanced cellular iron export despite the presence of hepcidin [36]. As discussed previously, activation of IL6/STAT3 pathway induces hepcidin transcription [24]. Therefore, targeting IL6/STAT3 pathway with anti-IL6 antibodies can down-regulate hepcidin expression. Siltuximab and tocilizumab, anti-IL6 antibody drugs, are in clinical use to reduce hepcidin effects and improve cellular iron export in patients with anemia [85]. Recent studies have shown that anti-IL6 antibody significantly decreases the expression of IL-6 protein in the brain of ischemic animal models [102]. Furthermore, anti-IL6 antibody rescues the blood–brain barrier from ischemic injury in the fetal sheep model [120]. In addition, urocortin, anti-inflammatory neuropeptide, suppresses inflammatory cytokine production (e.g., IL-6 and TNF-α) and reduces neurological deficits following ICH [63]. Taken together, targeting hepcidin and/or corresponding pathways could be an effective therapeutic approach to protect the brain from stroke injury.

6.7 Conclusion and Future Perspectives

Inflammation and iron accumulation play a crucial role in neuronal injuries caused by ischemic stroke. Hepcidin binds FPN1 and reduces its activity in excreting excess iron from the brain cells. The major factors that positively regulate hepcidin include iron status and inflammation. Modifying iron metabolism and post-stroke inflammation can reduce brain injury in stroke and the hepcidin pathway seems to be a promising molecular target for stroke therapy.

Increased hepcidin expression has emerged as a causative factor in several common iron disorders such as cerebral stroke. Targeting hepcidin pathways and/or interfering hepcidin–FPN1 interaction have become promising approaches for the development of novel therapeutics for iron disorders. However, multiple potential opportunities for cross communication among these pathways can cause off-target effects such as an increased risk of infection. For example, IL-6 receptor-targeted therapeutics impair host defense. Moreover, interfering hepcidin–FPN1 interaction by fursultiamine antagonizes hepcidin effects rather than reducing its expression. Therefore, lowering hepcidin expression directly in cerebral stroke where it is elevated may have less possible off-target effects in comparison with other therapeutic approaches mentioned previously. In addition, the direct targeting of hepcidin in stroke models would help us to understand the specific role of hepcidin in stroke-induced brain injury. The major challenge of systemic administration of anti-hepcidin is that some negative effects need to be taken into account. Deficiency of systemic hepcidin results in increased serum iron, dietary iron hyperabsorption, and iron-dependent oxidative stress. To overcome this challenge, anti-hepcidin could be delivered into the stroke-injured brain by using microinjection or nanoparticles as drug delivery systems. In the following, some experimental approaches are proposed in order to refine our understanding of the mechanistic role of hepcidin in cerebral stroke.

In summary, these future directions may improve our understating of the roles of hepcidin in iron-mediated neurotoxicity following cerebral stroke. Based on the results, novel therapeutic interventions might be developed to protect the brain from stroke injury.

Acknowledgements This investigation was supported by the University of Kansas General Research Fund. Mohammed M. A. Almutairi was supported by a scholarship from King Saud University (Riyadh, Saudi Arabia).

References

1. Abboud S, Haile DJ (2000) A novel mammalian iron-regulated protein involved in intracellular iron metabolism. J Biol Chem 275:19906–19912
2. Aguirre P, Mena N, Tapia V, Arredondo M, Nunez MT (2005) Iron homeostasis in neuronal cells: a role for IREG1. BMC Neurosci 6:3
3. Becerril-Ortega J, Bordji K, Freret T, Rush T, Buisson A (2014) Iron overload accelerates neuronal amyloid-beta production and cognitive impairment in transgenic mice model of Alzheimer's disease. Neurobiol Aging 35:2288–2301
4. Bedirli N, Bagriacik EU, Emmez H, Yilmaz G, Unal Y, Ozkose Z (2012) Sevoflurane and isoflurane preconditioning provides neuroprotection by inhibition of apoptosis-related mRNA expression in a rat model of focal cerebral ischemia. J Neurosurg Anesthesiol 24:336–344
5. Bergeron M, Gidday JM, Yu AY, Semenza GL, Ferriero DM, Sharp FR (2000) Role of hypoxia-inducible factor-1 in hypoxia-induced ischemic tolerance in neonatal rat brain. Ann Neurol 48:285–296
6. Bilotta F, Qeva E, Matot I (2016) Anesthesia and cognitive disorders: a systematic review of the clinical evidence. Expert Rev Neurother 16:1311–1320
7. Boserup MW, Lichota J, Haile D, Moos T (2011) Heterogenous distribution of ferroportin-containing neurons in mouse brain. Biometals 24:357–375
8. Boutin H, Lefeuvre RA, Horai R, Asano M, Iwakura Y, Rothwell NJ (2001) Role of IL-1alpha and IL-1beta in ischemic brain damage. J Neurosci 21:5528–5534
9. Burdo JR, Menzies SL, Simpson IA, Garrick LM, Garrick MD, Dolan KG, Haile DJ, Beard JL, Connor JR (2001) Distribution of divalent metal transporter 1 and metal transport protein 1 in the normal and Belgrade rat. J Neurosci Res 66:1198–1207
10. Cao S, Hua Y, Keep RF, Chaudhary N, Xi G (2018) Minocycline effects on intracerebral hemorrhage-induced iron overload in aged rats: brain iron quantification with magnetic resonance imaging. Stroke 49:995–1002
11. Carbonell T, Rama R (2007) Iron, oxidative stress and early neurological deterioration in ischemic stroke. Curr Med Chem 14:857–874
12. Castellanos M, Puig N, Carbonell T, Castillo J, Martinez J, Rama R, Davalos A (2002) Iron intake increases infarct volume after permanent middle cerebral artery occlusion in rats. Brain Res 952:1–6
13. Cekanaviciute E, Fathali N, Doyle KP, Williams AM, Han J, Buckwalter MS (2014) Astrocytic transforming growth factor-beta signaling reduces subacute neuroinflammation after stroke in mice. Glia 62:1227–1240
14. Chamorro A (2018) Neuroprotectants in the era of reperfusion therapy. J Stroke 20:197–207
15. Chen QX, Song SW, Chen QH, Zeng CL, Zheng X, Wang JL, Fang XM (2014) Silencing airway epithelial cell-derived hepcidin exacerbates sepsis induced acute lung injury. Crit Care 18:470
16. Chen Z, Gao C, Hua Y, Keep RF, Muraszko K, Xi G (2011) Role of iron in brain injury after intraventricular hemorrhage. Stroke 42:465–470
17. Clark RJ, Tan CC, Preza GC, Nemeth E, Ganz T, Craik DJ (2011) Understanding the structure/activity relationships of the iron regulatory peptide hepcidin. Chem Biol 18:336–343
18. Dai S, Hua Y, Keep RF, Novakovic N, Fei Z, Xi G (2018) Minocycline attenuates brain injury and iron overload after intracerebral hemorrhage in aged female rats. Neurobiol Dis
19. de Domenico I, Ward DM, Kaplan J (2011) Hepcidin and ferroportin: the new players in iron metabolism. Semin Liver Dis 31:272–279
20. de Domenico I, Ward DM, Langelier C, Vaughn MB, Nemeth E, Sundquist WI, Ganz T, Musci G, Kaplan J (2007) The molecular mechanism of hepcidin-mediated ferroportin down-regulation. Mol Biol Cell 18:2569–2578

21. Deb P, Sharma S, Hassan KM (2010) Pathophysiologic mechanisms of acute ischemic stroke: an overview with emphasis on therapeutic significance beyond thrombolysis. Pathophysiology 17:197–218
22. Deiner S, Silverstein JH (2009) Postoperative delirium and cognitive dysfunction. Br J Anaesth 103(Suppl 1):i41–i46
23. Dimitrijevic OB, Stamatovic SM, Keep RF, Andjelkovic AV (2007) Absence of the chemokine receptor CCR23 protects against cerebral ischemia/reperfusion injury in mice. Stroke 38:1345–1353
24. Ding H, Yan CZ, Shi H, Zhao YS, Chang SY, Yu P, Wu WS, Zhao CY, Chang YZ, Duan XL (2011) Hepcidin is involved in iron regulation in the ischemic brain. PLoS ONE 6:e25324
25. Dirnagl U, Iadecola C, Moskowitz MA (1999) Pathobiology of ischaemic stroke: an integrated view. Trends Neurosci 22:391–397
26. Donnan GA, Fisher M, Macleod M, Davis SM (2008) Stroke. Lancet 371:1612–1623
27. Donovan A, Brownlie A, Zhou Y, Shepard J, Pratt SJ, Moynihan J, Paw BH, Drejer A, Barut B, Zapata A, Law TC, Brugnara C, Lux SE, Pinkus GS, Pinkus JL, Kingsley PD, Palis J, Fleming MD, Andrews NC, Zon LI (2000) Positional cloning of zebrafish ferroportin1 identifies a conserved vertebrate iron exporter. Nature 403:776–781
28. Donovan A, Lima CA, Pinkus JL, Pinkus GS, Zon LI, Robine S, Andrews NC (2005) The iron exporter ferroportin/Slc40a1 is essential for iron homeostasis. Cell Metab 1:191–200
29. Ekdahl CT, Kokaia Z, Lindvall O (2009) Brain inflammation and adult neurogenesis: the dual role of microglia. Neuroscience 158:1021–1029
30. Fang MC, Go AS, Chang Y, Hylek EM, Henault LE, Jensvold NG, Singer DE (2007) Death and disability from warfarin-associated intracranial and extracranial hemorrhages. Am J Med 120:700–705
31. Feng Y, Leblanc MH, Leblanc EB, Parker CC, Fratkin JD, Qian XB, Patel DM, Huang M, Smith EE, Vig PJ (2000) Desmethyl tirilazad improves neurologic function after hypoxic ischemic brain injury in piglets. Crit Care Med 28:1431–1438
32. Fernandes A, Preza GC, Phung Y, de Domenico I, Kaplan J, Ganz T, Nemeth E (2009) The molecular basis of hepcidin-resistant hereditary hemochromatosis. Blood 114:437–443
33. Fleming RE, Sly WS (2002) Mechanisms of iron accumulation in hereditary hemochromatosis. Annu Rev Physiol 64:663–680
34. Fogarty Mack P (2014) Intracranial haemorrhage: therapeutic interventions and anaesthetic management. Br J Anaesth 113(Suppl 2):ii17–ii25
35. Fujita N, Sugimoto R, Takeo M, Urawa N, Mifuji R, Tanaka H, Kobayashi Y, Iwasa M, Watanabe S, Adachi Y, Kaito M (2007) Hepcidin expression in the liver: relatively low level in patients with chronic hepatitis C. Mol Med 13:97–104
36. Fung E, Sugianto P, Hsu J, Damoiseaux R, Ganz T, Nemeth E (2013) High-throughput screening of small molecules identifies hepcidin antagonists. Mol Pharmacol 83:681–690
37. Ganz T (2005) Cellular iron: ferroportin is the only way out. Cell Metab 1:155–157
38. Ganz T (2005) Hepcidin–a regulator of intestinal iron absorption and iron recycling by macrophages. Best Pract Res Clin Haematol 18:171–182
39. Ganz T, Nemeth E (2012) Hepcidin and iron homeostasis. Biochim Biophys Acta 1823:1434–1443
40. Garcia-Rodriguez LA, Gaist D, Morton J, Cookson C, Gonzalez-Perez A (2013) Antithrombotic drugs and risk of hemorrhagic stroke in the general population. Neurology 81:566–574
41. Garcia-Yebenes I, Sobrado M, Moraga A, Zarruk JG, Romera VG, Pradillo JM, de la Ossa NP, Moro MA, Davalos A, Lizasoain I (2012) Iron overload, measured as serum ferritin, increases brain damage induced by focal ischemia and early reperfusion. Neurochem Int 61:1364–1369
42. Garcia JM, Stillings SA, Leclerc JL, Phillips H, Edwards NJ, Robicsek SA, Hoh BL, Blackburn S, Dore S (2017) Role of interleukin-10 in acute brain injuries. Front Neurol 8:244
43. Garton T, Keep RF, Hua Y, Xi G (2016) Brain iron overload following intracranial haemorrhage. Stroke Vasc Neurol 1:172–184

44. Go AS, Mozaffarian D, Roger VL, Benjamin EJ, Berry JD, Blaha MJ, Dai S, Ford ES, Fox CS, Franco S, Fullerton HJ, Gillespie C, Hailpern SM, Heit JA, Howard VJ, Huffman MD, Judd SE, Kissela BM, Kittner SJ, Lackland DT, Lichtman JH, Lisabeth LD, Mackey RH, Magid DJ, Marcus GM, Marelli A, Matchar DB, McGuire DK, Mohler ER 3rd, Moy CS, Mussolino ME, Neumar RW, Nichol G, Pandey DK, Paynter NP, Reeves MJ, Sorlie PD, Stein J, Towfighi A, Turan TN, Virani SS, Wong ND, Woo D, Turner MB (2014) Heart disease and stroke statistics–2014 update: a report from the American Heart Association. Circulation 129:e28–e292
45. Guo C, Wang P, Zhong ML, Wang T, Huang XS, Li JY, Wang ZY (2013) Deferoxamine inhibits iron induced hippocampal tau phosphorylation in the Alzheimer transgenic mouse brain. Neurochem Int 62:165–172
46. Guo P, Cui R, Chang YZ, Wu WS, Qian ZM, Yoshida K, Qiao YT, Takeda S, Duan XL (2009) Hepcidin, an antimicrobial peptide is downregulated in ceruloplasmin-deficient mice. Peptides 30:262–266
47. Hall ED (1992) Novel inhibitors of iron-dependent lipid peroxidation for neurodegenerative disorders. Ann Neurol 32(Suppl):S137–S142
48. Hanson LR, Roeytenberg A, Martinez PM, Coppes VG, Sweet DC, Rao RJ, Marti DL, Hoekman JD, Matthews RB, Frey WH 2nd, Panter SS (2009) Intranasal deferoxamine provides increased brain exposure and significant protection in rat ischemic stroke. J Pharmacol Exp Ther 330:679–686
49. Hemorrhagic Stroke Academia Industry Roundtable Participants (2018) Unmet needs and challenges in clinical research of intracerebral hemorrhage. Stroke 49:1299–1307
50. Hentze MW, Muckenthaler MU, Andrews NC (2004) Balancing acts: molecular control of mammalian iron metabolism. Cell 117:285–297
51. Hentze MW, Muckenthaler MU, Galy B, Camaschella C (2010) Two to tango: regulation of Mammalian iron metabolism. Cell 142:24–38
52. Hishikawa T, Ono S, Ogawa T, Tokunaga K, Sugiu K, Date I (2008) Effects of deferoxamine-activated hypoxia-inducible factor-1 on the brainstem after subarachnoid hemorrhage in rats. Neurosurgery 62:232–240 (discussion 240-1)
53. Hua Y, Keep RF, Hoff JT, Xi G (2007) Brain injury after intracerebral hemorrhage: the role of thrombin and iron. Stroke 38:759–762
54. Jomova K, Valko M (2011) Advances in metal-induced oxidative stress and human disease. Toxicology 283:65–87
55. Kanda J, Mizumoto C, Kawabata H, Tsuchida H, Tomosugi N, Matsuo K, Uchiyama T (2008) Serum hepcidin level and erythropoietic activity after hematopoietic stem cell transplantation. Haematologica 93:1550–1554
56. Kidwell CS, Saver JL, Starkman S, Duckwiler G, Jahan R, Vespa P, Villablanca JP, Liebeskind DS, Gobin YP, Vinuela F, Alger JR (2002) Late secondary ischemic injury in patients receiving intraarterial thrombolysis. Ann Neurol 52:698–703
57. Kim JS, Gautam SC, Chopp M, Zaloga C, Jones ML, Ward PA, Welch KM (1995) Expression of monocyte chemoattractant protein-1 and macrophage inflammatory protein-1 after focal cerebral ischemia in the rat. J Neuroimmunol 56:127–134
58. Kondo Y, Ogawa N, Asanuma M, Ota Z, Mori A (1995) Regional differences in late-onset iron deposition, ferritin, transferrin, astrocyte proliferation, and microglial activation after transient forebrain ischemia in rat brain. J Cereb Blood Flow Metab 15:216–226
59. Lakhan SE, Kirchgessner A, Hofer M (2009) Inflammatory mechanisms in ischemic stroke: therapeutic approaches. J Transl Med 7:97
60. Lampl Y, Boaz M, Gilad R, Lorberboym M, Dabby R, Rapoport A, Anca-Hershkowitz M, Sadeh M (2007) Minocycline treatment in acute stroke: an open-label, evaluator-blinded study. Neurology 69:1404–1410
61. Lee P, Peng H, Gelbart T, Wang L, Beutler E (2005) Regulation of hepcidin transcription by interleukin-1 and interleukin-6. Proc Natl Acad Sci U S A 102:1906–1910

62. Li L, Li YW, Zhao JY, Liu YZ, Holscher C (2009) Quantitative analysis of iron concentration and expression of ferroportin 1 in the cortex and hippocampus of rats induced by cerebral ischemia. J Clin Neurosci 16:1466–1472
63. Liew HK, Pang CY, Hsu CW, Wang MJ, Li TY, Peng HF, Kuo JS, Wang JY (2012) Systemic administration of urocortin after intracerebral hemorrhage reduces neurological deficits and neuroinflammation in rats. J Neuroinflammation 9:13
64. Lipscomb DC, Gorman LG, Traystman RJ, Hurn PD (1998) Low molecular weight iron in cerebral ischemic acidosis in vivo. Stroke 29:487–492 (discussion 493)
65. Lucas SM, Rothwell NJ, Gibson RM (2006) The role of inflammation in CNS injury and disease. Br J Pharmacol 147(Suppl 1):S232–S240
66. Mason SE, Noel-Storr A, Ritchie CW (2010) The impact of general and regional anesthesia on the incidence of post-operative cognitive dysfunction and post-operative delirium: a systematic review with meta-analysis. J Alzheimers Dis 22(Suppl 3):67–79
67. McKie AT, Marciani P, Rolfs A, Brennan K, Wehr K, Barrow D, Miret S, Bomford A, Peters TJ, Farzaneh F, Hediger MA, Hentze MW, Simpson RJ (2000) A novel duodenal iron-regulated transporter, IREG1, implicated in the basolateral transfer of iron to the circulation. Mol Cell 5:299–309
68. Merle U, Fein E, Gehrke SG, Stremmel W, Kulaksiz H (2007) The iron regulatory peptide hepcidin is expressed in the heart and regulated by hypoxia and inflammation. Endocrinology 148:2663–2668
69. Monk TG, Weldon BC, Garvan CW, Dede DE, van der Aa MT, Heilman KM, Gravenstein JS (2008) Predictors of cognitive dysfunction after major noncardiac surgery. Anesthesiology 108:18–30
70. Moos T, Rosengren Nielsen T, Skjorringe T, Morgan EH (2007) Iron trafficking inside the brain. J Neurochem 103:1730–1740
71. Nakamura T, Keep RF, Hua Y, Nagao S, Hoff JT, Xi G (2006) Iron-induced oxidative brain injury after experimental intracerebral hemorrhage. Acta Neurochir Suppl 96:194–198
72. Nemeth E, Ganz T (2006) Regulation of iron metabolism by hepcidin. Annu Rev Nutr 26:323–342
73. Nemeth E, Rivera S, Gabayan V, Keller C, Taudorf S, Pedersen BK, Ganz T (2004) IL-6 mediates hypoferremia of inflammation by inducing the synthesis of the iron regulatory hormone hepcidin. J Clin Invest 113:1271–1276
74. Nemeth E, Tuttle MS, Powelson J, Vaughn MB, Donovan A, Ward DM, Ganz T, Kaplan J (2004) Hepcidin regulates cellular iron efflux by binding to ferroportin and inducing its internalization. Science 306:2090–2093
75. Ni W, Okauchi M, Hatakeyama T, Gu Y, Keep RF, Xi G, Hua Y (2015) Deferoxamine reduces intracerebral hemorrhage-induced white matter damage in aged rats. Exp Neurol 272:128–134
76. Nicolas G, Bennoun M, Devaux I, Beaumont C, Grandchamp B, Kahn A, Vaulont S (2001) Lack of hepcidin gene expression and severe tissue iron overload in upstream stimulatory factor 2 (USF2) knockout mice. Proc Natl Acad Sci U S A 98:8780–8785
77. Nicolas G, Viatte L, Lou DQ, Bennoun M, Beaumont C, Kahn A, Andrews NC, Vaulont S (2003) Constitutive hepcidin expression prevents iron overload in a mouse model of hemochromatosis. Nat Genet 34:97–101
78. Ooboshi H, Ibayashi S, Shichita T, Kumai Y, Takada J, Ago T, Arakawa S, Sugimori H, Kamouchi M, Kitazono T, Iida M (2005) Postischemic gene transfer of interleukin-10 protects against both focal and global brain ischemia. Circulation 111:913–919
79. Palmer C, Menzies SL, Roberts RL, Pavlick G, Connor JR (1999) Changes in iron histochemistry after hypoxic-ischemic brain injury in the neonatal rat. J Neurosci Res 56:60–71
80. Palmer C, Roberts RL, Bero C (1994) Deferoxamine posttreatment reduces ischemic brain injury in neonatal rats. Stroke 25:1039–1045
81. Patt A, Horesh IR, Berger EM, Harken AH, Repine JE (1990) Iron depletion or chelation reduces ischemia/reperfusion-induced edema in gerbil brains. J Pediatr Surg 25:224–227 (discussion 227-8)

82. Perna R, Temple J (2015) Rehabilitation outcomes: ischemic versus hemorrhagic strokes. Behav Neurol 2015:891651
83. Pigeon C, Ilyin G, Courselaud B, Leroyer P, Turlin B, Brissot P, Loreal O (2001) A new mouse liver-specific gene, encoding a protein homologous to human antimicrobial peptide hepcidin, is overexpressed during iron overload. J Biol Chem 276:7811–7819
84. Poggiali E, Cassinerio E, Zanaboni L, Cappellini MD (2012) An update on iron chelation therapy. Blood Transfus 10:411–422
85. Poli M, Asperti M, Ruzzenenti P, Regoni M, Arosio P (2014) Hepcidin antagonists for potential treatments of disorders with hepcidin excess. Front Pharmacol 5:86
86. Prass K, Ruscher K, Karsch M, Isaev N, Megow D, Priller J, Scharff A, Dirnagl U, Meisel A (2002) Desferrioxamine induces delayed tolerance against cerebral ischemia in vivo and in vitro. J Cereb Blood Flow Metab 22:520–525
87. Qian ZM, Shen X (2001) Brain iron transport and neurodegeneration. Trends Mol Med 7:103–108
88. Righy C, Turon R, Freitas G, Japiassu AM, Faria Neto HCC, Bozza M, Oliveira MF, Bozza FA (2018) Hemoglobin metabolism by-products are associated with an inflammatory response in patients with hemorrhagic stroke. Rev Bras Ter Intensiva 30:21–27
89. Roetto A, Papanikolaou G, Politou M, Alberti F, Girelli D, Christakis J, Loukopoulos D, Camaschella C (2003) Mutant antimicrobial peptide hepcidin is associated with severe juvenile hemochromatosis. Nat Genet 33:21–22
90. Rosenthal RE, Chanderbhan R, Marshall G, Fiskum G (1992) Prevention of post-ischemic brain lipid conjugated diene production and neurological injury by hydroxyethyl starch-conjugated deferoxamine. Free Radic Biol Med 12:29–33
91. Roth JM (2011) Recombinant tissue plasminogen activator for the treatment of acute ischemic stroke. Proc (Bayl Univ Med Cent) 24:257–259
92. Saleppico S, Mazzolla R, Boelaert JR, Puliti M, Barluzzi R, Bistoni F, Blasi E (1996) Iron regulates microglial cell-mediated secretory and effector functions. Cell Immunol 170:251–259
93. Selim M (2010) Treatment with the iron chelator, deferoxamine mesylate, alters serum markers of oxidative stress in stroke patients. Transl Stroke Res 1:35–39
94. Selim MH, Ratan RR (2004) The role of iron neurotoxicity in ischemic stroke. Ageing Res Rev 3:345–353
95. Seymour DG, Severn AM (2009) Cognitive dysfunction after surgery and anaesthesia: what can we tell the grandparents? Age Ageing 38:147–150
96. Shichita T, Ito M, Yoshimura A (2014) Post-ischemic inflammation regulates neural damage and protection. Front Cell Neurosci 8:319
97. Soloniuk DS, Perkins E, Wilson JR (1992) Use of allopurinol and deferoxamine in cellular protection during ischemia. Surg Neurol 38:110–113
98. Soriano SG, Amaravadi LS, Wang YF, Zhou H, Yu GX, Tonra JR, Fairchild-Huntress V, Fang Q, Dunmore JH, Huszar D, Pan Y (2002) Mice deficient in fractalkine are less susceptible to cerebral ischemia-reperfusion injury. J Neuroimmunol 125:59–65
99. Sorond FA, Tan CO, Larose S, Monk AD, Fichorova R, Ryan S, Lipsitz LA (2015) Deferoxamine, cerebrovascular hemodynamics, and vascular aging: potential role for hypoxia-inducible transcription factor-1-regulated pathways. Stroke 46:2576–2583
100. Steinmetz J, Christensen KB, Lund T, Lohse N, Rasmussen LS, Group I (2009) Long-term consequences of postoperative cognitive dysfunction. Anesthesiology 110:548–555
101. Tan G, Liu L, He Z, Sun J, Xing W, Sun X (2016) Role of hepcidin and its downstream proteins in early brain injury after experimental subarachnoid hemorrhage in rats. Mol Cell Biochem 418:31–38
102. Tang YH, Vital S, Russell J, Seifert H, Neil Granger D (2015) Interleukin-6 mediates enhanced thrombus development in cerebral arterioles following a brief period of focal brain ischemia. Exp Neurol 271:351–357
103. Taylor RA, Chang CF, Goods BA, Hammond MD, Mac Grory B, Ai Y, Steinschneider AF, Renfroe SC, Askenase MH, McCullough LD, Kasner SE, Mullen MT, Hafler DA, Love JC, Sansing LH (2017) TGF-beta1 modulates microglial phenotype and promotes recovery after intracerebral hemorrhage. J Clin Invest 127:280–292

104. Tobin MK, Bonds JA, Minshall RD, Pelligrino DA, Testai FD, Lazarov O (2014) Neurogenesis and inflammation after ischemic stroke: what is known and where we go from here. J Cereb Blood Flow Metab 34:1573–1584
105. Urrutia P, Aguirre P, Esparza A, Tapia V, Mena NP, Arredondo M, Gonzalez-Billault C, Nunez MT (2013) Inflammation alters the expression of DMT1, FPN1 and hepcidin, and it causes iron accumulation in central nervous system cells. J Neurochem 126:541–549
106. Verga Falzacappa MV, Vujicspasic M, Kessler R, Stolte J, Hentze MW, Muckenthaler MU (2007) STAT3 mediates hepatic hepcidin expression and its inflammatory stimulation. Blood 109:353–358
107. Wang G, Hu W, Tang Q, Wang L, Sun XG, Chen Y, Yin Y, Xue F, Sun Z (2016) Effect comparison of both iron chelators on outcomes, iron deposit, and iron transporters after intracerebral hemorrhage in rats. Mol Neurobiol 53:3576–3585
108. Wang Q, Du F, Qian ZM, Ge XH, Zhu L, Yung WH, Yang L, Ke Y (2008) Lipopolysaccharide induces a significant increase in expression of iron regulatory hormone hepcidin in the cortex and substantia nigra in rat brain. Endocrinology 149:3920–3925
109. Wang SM, Fu LJ, Duan XL, Crooks DR, Yu P, Qian ZM, Di XJ, Li J, Rouault TA, Chang YZ (2010) Role of hepcidin in murine brain iron metabolism. Cell Mol Life Sci 67:123–133
110. Ward RJ, Zucca FA, Duyn JH, Crichton RR, Zecca L (2014) The role of iron in brain ageing and neurodegenerative disorders. Lancet Neurol 13:1045–1060
111. Wilkinson DA, Pandey AS, Thompson BG, Keep RF, Hua Y, Xi G (2018) Injury mechanisms in acute intracerebral hemorrhage. Neuropharmacology 134:240–248
112. Wrighting DM, Andrews NC (2006) Interleukin-6 induces hepcidin expression through STAT3. Blood 108:3204–3209
113. Wu H, Wu T, Xu X, Wang J, Wang J (2011) Iron toxicity in mice with collagenase-induced intracerebral hemorrhage. J Cereb Blood Flow Metab 31:1243–1250
114. Wu J, Hua Y, Keep RF, Schallert T, Hoff JT, Xi G (2002) Oxidative brain injury from extravasated erythrocytes after intracerebral hemorrhage. Brain Res 953:45–52
115. Xiao L, Li XQ, Zhang HX (2004) Effects of naoyian granule on the expression of IL-6 mRNA and protein in the experimental intracerebral hemorrhagic brain in rats. Zhong Nan Da Xue Xue Bao Yi Xue Ban 29:177–180
116. Xie J, Brayne C, Matthews FE, Medical Research Council Cognitive F, Ageing Study C (2008) Survival times in people with dementia: analysis from population based cohort study with 14 year follow-up. BMJ 336:258–262
117. Xiong XY, Liu L, Wang FX, Yang YR, Hao JW, Wang PF, Zhong Q, Zhou K, Xiong A, Zhu WY, Zhao T, Meng ZY, Wang YC, Gong QW, Liao MF, Wang J, Yang QW (2016) Toll-like receptor 4/MyD88-mediated signaling of hepcidin expression causing brain iron accumulation, oxidative injury, and cognitive impairment after intracerebral hemorrhage. Circulation 134:1025–1038
118. Yamasaki Y, Matsuura N, Shozuhara H, Onodera H, Itoyama Y, Kogure K (1995) Interleukin-1 as a pathogenetic mediator of ischemic brain damage in rats. Stroke 26:676–680 (discussion 681)
119. Yang L, Zhang B, Yin L, Cai B, Shan H, Zhang L, Lu Y, Bi Z (2011) Tanshinone IIA prevented brain iron dyshomeostasis in cerebral ischemic rats. Cell Physiol Biochem 27:23–30
120. Zhang J, Sadowska GB, Chen X, Park SY, Kim JE, Bodge CA, Cummings E, Lim YP, Makeyev O, Besio WG, Gaitanis J, Banks WA, Stonestreet BS (2015) Anti-IL-6 neutralizing antibody modulates blood-brain barrier function in the ovine fetus. FASEB J 29:1739–1753
121. Zhang XY, Cao JB, Zhang LM, Li YF, Mi WD (2015) Deferoxamine attenuates lipopolysaccharide-induced neuroinflammation and memory impairment in mice. J Neuroinflammation 12:20
122. Zhao Y, Rempe DA (2011) Prophylactic neuroprotection against stroke: low-dose, prolonged treatment with deferoxamine or deferasirox establishes prolonged neuroprotection independent of HIF-1 function. J Cereb Blood Flow Metab 31:1412–1423
123. Zheng Z, Yenari MA (2004) Post-ischemic inflammation: molecular mechanisms and therapeutic implications. Neurol Res 26:884–892

124. Zhou Y, Wang Y, Wang J, Anne Stetler R, Yang QW (2014) Inflammation in intracerebral hemorrhage: from mechanisms to clinical translation. Prog Neurobiol 115:25–44
125. Zohn IE, de Domenico I, Pollock A, Ward DM, Goodman JF, Liang X, Sanchez AJ, Niswander L, Kaplan J (2007) The flatiron mutation in mouse ferroportin acts as a dominant negative to cause ferroportin disease. Blood 109:4174–4180

Chapter 7
Iron Pathophysiology in Friedreich's Ataxia

Kuanyu Li

Abstract Friedreich's ataxia (FRDA) is a degenerative disease that affects both the central and the peripheral nervous systems and non-neural tissues including, mainly, heart, and endocrine pancreas. It is an autosomal recessive disease caused by a GAA triplet-repeat localized within an Alu sequence element in intron 1 of frataxin (*FXN*) gene, which encodes a mitochondrial protein FXN. This protein is essential for mitochondrial function by the involvement of iron–sulfur cluster biogenesis. The effects of its deficiency also include disruption of cellular, particularly mitochondrial, iron homeostasis, i.e., relatively more iron accumulated in mitochondria and less iron presented in cytosol. Though iron toxicity is commonly thought to be mediated via Fenton reaction, oxidative stress seems not to be the main problem to result in detrimental effects on cell survival, particularly neuron survival. Therefore, the basic research on FXN function is urgently demanded to understand the disease. This chapter focuses on the outcome of FXN expression, regulation, and function in cellular or animal models of FRDA and on iron pathophysiology in the affected tissues. Finally, therapeutic strategies based on the control of iron toxicity and iron cellular redistribution are considered. The combination of multiple therapeutic targets including iron, oxidative stress, mitochondrial function, and FXN regulation is also proposed.

Keywords Friedreich ataxia · Frataxin · Iron · Iron–sulfur cluster biogenesis

7.1 Friedreich's Ataxia (FRDA)

FRDA, described by Nikolaus Friedreich in 1863, is a single-gene, recessive, and progressive neurodegenerative disorder that classically appears around puberty, rarely in early childhood, and in some cases later in life. On average, in ten or fifteen years after the onset of the disease, patients lose the ability to walk, to stand or to sit up

K. Li (✉)
Jiangsu Key Laboratory of Molecular Medicine, Medical School of Nanjing University, Nanjing 210093, People's Republic of China
e-mail: likuanyu@nju.edu.cn

Y.-Z. Chang (ed.), *Brain Iron Metabolism and CNS Diseases*,
Advances in Experimental Medicine and Biology 1173,
https://doi.org/10.1007/978-981-13-9589-5_7

unaided. The disease occurs mainly in Caucasians. It is believed that the restricted distribution in Caucasians is due to a single founder mutation that occurred between 500 and 800 generations ago [1].

The essential clinical and pathological features of the disease include the degeneration in the spinal cord of the posterior columns, which contain axons of dorsal-root-ganglia sensory neurons. Most patients succumb to cardiomyopathy, and many become diabetic during the course of the disease. Typically, cognitive functions are preserved well. The presence of comorbid depression in FRDA is recently recognized [2]. In 1996, came the breakthrough identification of the disease gene, frataxin (*FXN*), for FRDA [3].

The last 20 years have seen remarkable progress in elucidating the function of the encoded protein, FXN, which localizes to the mitochondria [4] and is involved in iron-sulfur cluster biogenesis [5]. The focus of this chapter is on iron dysregulation in FRDA, with a particular emphasis on the most recent findings.

7.2 *FXN* Gene and Its Function

FXN is a highly conserved gene present from gram-negative bacteria to eukaryotes, including yeast and mammals. In human, *FXN* maps on chromosome 9, at 9q13–q21.1. The prevalent mutation is an abnormal expansion of a GAA triplet-repeat localized within an Alu sequence element in intron 1, which accounts for greater than 98% cases of mutation in patients. Homozygotes (bi-allelic expansion) are found in the majority (96–98%) of FRDA patients, and the remaining ones are compound of heterozygotes of the (GAA)n expansion and a point mutation (missense, nonsense or splicing) [4, 5] or an intragenic deletion [6] or intronic deletion [7].

7.2.1 FXN *Gene Mutation and Its Expression of Isoforms*

Ages of onset and death as a function of GAA trinucleotide repeat expansions are closely correlated with expression levels of *FXN*. This GAA expansion ranges from fewer than 200 to as many as 1500 repeats, in which size of the expansion correlates negatively with expression of the encoded protein and positively with the ages of onset and severity of the disease. This expansion leads to transcriptional silencing of *FXN* through a commonly accepted mechanism involving modifications of the chromatin structure of the locus, resulting in expression of a structurally and functionally normal FXN but at levels that are estimated at 5–30% of normal [8, 9]. The decreased histone acetylation and extensive methylation of CpG regions upstream of the GAA repeats are observed in patient cell lines and tissues [10, 11], suggesting that the enhanced heterochromatin formation might impede the transcription of *FXN*, leading to the lower levels of FXN protein [12–14]. Studies employing an experimental histone deacetylase inhibitor (HDACi) in a mouse model (KIKI with GAA insertion) of FRDA revealed that this drug can substantially increase FXN mRNA and protein levels [13]. Unexpectedly, the phase III clinical trial did not prove the effects of HDACi on clinical symptoms.

Reduction of *FXN* transcription may result from reduced accessibility of transcriptional regulatory factors to the promoter region and/or trinucleotide repeat region [12]. However, the identity and number of the regulatory factors influencing *FXN* expression are largely unknown. The known human transcription factors have been proposed such as SRF and AP2 [15], HIF1 [16], and p53 [17] to regulate *FXN* expression. The entity of the transcriptional regulatory machinery involved in *FXN* expression needs to be further investigated to aid the identification of drugs or therapies directed at restoring FXN protein levels in affected tissues of FRDA.

FXN is synthesized as a precursor protein (210 amino acids, FXN1–210) that is then matured in two steps to give an intermediate form (FXN42–210) and the major mature form (FXN81–210) in the mitochondrial matrix [15, 18]. A few studies support the presence of the extra-mitochondrial isoforms [19–22], and the origin of the extra-mitochondrial FXN is proposed [21, 22] although the functional rational is controversial [22, 23]. More isoforms, e.g., in erythrocytes, cerebellum, and heart, are revealed [21, 22, 24], and increasing the abundance of the heart-specific isoform III significantly increased the mitochondrial aconitase activity, while over-expression of the cerebellum-specific isoform II protected against oxidative damage of Fe–S cluster-containing aconitase [22]. More specifically, monomeric FXN81–210 donates Fe(2+) for Fe–S cluster assembly on ISCU, whereas oligomeric FXN42–210 donates either Fe(2+) or Fe(3+) based on important functional differences observed between FXN81–210 and oligomeric FXN42–210 [25]. In the same study, authors even proposed that isoform FXN81-210 takes a role in steady stage, while FXN42-210 does under conditions of increased Fe-S cluster demand. More work needs to be done to understand the mechanism of tissue-specific pathology in FRDA.

7.2.2 FXN Function

FXN is mainly localized within the eukaryotic mitochondria and is ubiquitously expressed in mammals. The structure of FXN is unique and conserved between species: a small globular acidic protein composed of a long N-terminal alpha helix and a short C-terminal alpha helix that both interact with a central beta-sheet structure [26, 27].

To understand the role and pathophysiological implication of iron in the disease, it is therefore essential to understand the function of FXN and how its deficit can lead to cellular iron dysregulation. Initially, iron got extensive attention because of the impaired activities of iron–sulfur cluster-containing enzymes including mitochondrial aconitase, respiratory complex I, II, and III in affected cells derived from FRDA patients [28]. The consequence is the dramatic reduction in ATP production. Hence, mitochondrial dysfunction was suspected of contributing to the pathophysiology of FRDA.

Though multiple functions of FXN have been proposed, a commonly accepted hypothesis is that FXN is primarily involved in Fe–S cluster biogenesis. Because Fe–S cluster is a very important prosthetic group fulfilling functions in electron transport, enzyme catalysis, homeostatic regulation, and sulfur activation, FXN deficiency-induced Fe–S cluster deficits reduce the enzymatic activities of Fe–S proteins in all eukaryotes [29]. In order to investigate the exact function of FXN, interactants were revealed with different methods. However, data from mammals, yeast, and bacteria were quite controversial as to the direct FXN protein partner in Fe–S biogenesis [30–33]. The only interaction partners of FXN have been extensively and convincingly identified to be NFS1(+ISC11) and ISCU, which complex is considered to be the core components of Fe–S cluster biosynthesis machinery (reviewed in [29]).

7.2.2.1 Iron Chaperone and Iron Storage

Studies suggested that FXN might be the iron donor or iron chaperone for the assembly of the Fe–S cluster [25, 34–36]. It was proposed in early studies that FXN could act as a metal chaperone because the structure indicated the iron-binding sites of FXN [37] and the purified protein could coordinate iron [38]. There is also uncertainty on the number of metal ions that are bound per FXN monomer, whereas 2 ferrous iron or 6–7 ferric iron ions per FXN monomer were suggested using a group of exposed acidic residues [37, 38]. These amino acids are different from the canonical iron-binding motifs, which normally are cysteine and/or histidine. The rational for the iron-binding activity of FXN was proposed by the interaction between FXN and ISCU [30] and between FXN and ferrochelatase (FECH) [39]. The former complex facilitates the transfer of iron from holo-FXN to nucleation sites for [2Fe–2S] cluster formation on ISU [38], while the latter complex mediates iron insertion into protoporphyrin IX to make heme [40]. The destination of iron delivery by FXN was determined by the cellular physiological activities, i.e., mainly to FECH for heme biosynthesis in erythrocytes by down-regulation of FXN and to both ISCU and FECH in normal cells to fulfill the two major iron-dependent pathways [40]. However, a recent study demonstrated that no changes in heme synthesis was observed in human FRDA erythroid progenitor cells compared with healthy control cells, which was supported by that FRDA patients do not have abnormal hematological symptoms. As for the distinct high binding affinities of holo-FXN to the target protein FECH, the binding between might be a negative mechanism for removal of FXN during erythropoiesis. In non-erythrocytes, it is relevant that holo-FXN delivers iron to ISCU, particularly in cells with highly expressed FXN, for instance in FRDA-affected tissues.

More recently, Isaya and colleagues found that iron binding to in vitro purified FXN promoted its oligomerization to form a complex of high molecular weight (850–1100 kDa) [42, 43]. These oligomeric forms resemble ferritin structure containing thousands of iron ions for iron storage [42, 44]. Therefore, FXN is proposed to act as ferritin for mitochondrial iron storage. Indeed, the ferroxidase activity of FXN was revealed in yeast [45, 46], further supporting the iron-binding activity. However, another study revealed that human mitochondrial ferritin improves respiratory function in yeast mutants deficient in iron–sulfur cluster biogenesis, but is not a functional homologue of yeast FXN [47]. Concerning the iron-binding activity and iron-induced oligomerization, more detailed investigation has elucidated that a $[FXN42–210]_{24}$:$[NFS1]_{24}$:$[ISD11]_{24}$:$[ISCU]_{24}$ complex is consistent with the measured 1:1:1:1 stoichiometry of its four components [35]. This structure of the complex does not only recapitulate the known mechanism for sulfur donation from NFS1 to ISCU but also provides a path for iron donation from FXN42–210 to ISCU. However, isoforms FXN42–210 and FXN81–210 have strikingly different biochemical properties (reviewed in [48]). The former one has not yet accepted abundant in heart, cerebellum, and dividing lymphoblasts, and the later one only forms monomeric complex in vivo. Overall, it was suggested that the iron binding of the mature form is served for delivery of iron to biochemical pathways, while FXN42–210 could be able to oligomerize and store iron as ferritin.

7.2.2.2 Regulator of Cysteine Desulfurase in Fe–S Cluster Biogenesis

The role of FXN in orchestrating Fe–S cluster formation is widely accepted. Iron is considered first to be imported into the mitochondria by mitoferrins (MFRN1 and MFRN2) [49]. MFRN1 is inducibly expressed during erythropoiesis, and MFRN2 is ubiquitously expressed in all tissues, probably, for basal mitochondrial iron uptake. The iron is then majorly used for Fe–S cluster and heme biogenesis within mitochondria though how the iron was delivered into the process is in debate. De novo Fe–S cluster biogenesis involves assembly of inorganic sulfur and iron on a scaffold protein ISCU, with which a complex consists of NFS1-ISD11, the cysteine desulfurase providing the sulfur, and FXN. A few studies support that FXN regulates the NFS1 activity and the entry of iron within the complex via stabilizing the complex [50–52], which is in consistence with the observation that the binding of FXN with NFS1 is not iron dependent [23]. These results would not support FXN functions as an iron donor for Fe–S cluster biosynthesis, rather than as an allosteric regulator for enhancing NFS1 function to form persulfide on NFS1 and to transfer sulfur from NFS1 to ISCU as an initial step in Fe–S cluster biosynthesis [52]. Still, how FXN stabilizes the quaternary complex of core components and controls iron entry to the complex and how iron gets to the entry and is transferred to scaffold protein ISCU are mysterious.

7.2.2.3 Other Putative Functions

Though the functions of FXN have been proposed many, for instance, involvements against oxidative stress and heme biogenesis, they may be the secondary effects of

Fe–S cluster deficiency. The lack of FXN prevents the assembly of Fe–S clusters with consecutive mitochondrial iron overload [53]. The toxic radicals through the Fenton reaction are generated at an excessively high level in FXN-deficient mitochondria. Meanwhile, due to the lack of Fe–S clusters as important cofactors in complex I, II, and III of electron transport chain, the respiratory chain is defective, which makes reactive oxygen species (ROS) produced more severely. These ROS can oxidize ferrous iron and disrupt the cellular redox homeostasis. The consequence will be the damage of macromolecules including DNA, proteins, and lipids, which may be the pathological mechanism of FRDA. Interestingly, not all mitochondria-rich tissues are affected in FRDA patients, but clearly cerebellum, spinal cord, and heart.

Because ferrochelatase, the last enzyme for heme biogenesis, needs Fe–S cluster to gain the enzymatic activity for insertion of iron into protoporphyrin IX to form heme, the lack of Fe–S cluster induced by FXN deficiency would affect heme biogenesis. Therefore, FRDA patients would develop hematological disorder. FXN was found to interact with ferrochelatase, which hints that FXN delivers iron to ferrochelatase to produce heme in cells, particularly in erythroblasts. However, FRDA patients do not present typical hematological disorder [41], which suggests that FXN may also play a role in determining the recipient of this prosthetic group after assembly, favorably being transferred to mitochondrial respiratory complexes.

7.3 Working Models for FRDA Disease

The functions of FXN have been proposed involving biogenesis of iron–sulfur clusters, iron chaperoning, iron storage, and control of iron-mediated oxidative tissue damage (reviewed in [54]). As demonstrated in knockout animals, complete absence of *FXN* leads to early embryonic lethality [55], indicating that at least some FXN function is essential for survival. Despite the difficulty in generating perfect FRDA models, a multitude of complementary models have been generated enabling significant advances in understanding the function of FXN in model organisms as yeast, fruit fly, nematode, and mouse.

In yeast, the Yfh1 protein (encoded by yeast *FXN* homolog Yfh1p) is confirmed to be involved in iron homeostasis and mitochondrial function [56]. The deletion of the Yfh1 gene makes mitochondrial iron accumulation (up to 10 times more than the wild strain) makes the cells highly sensitive to oxidative stress, and also makes the strain unable to grow on a non-fermentable source of carbon [53, 56], suggesting that it could not carry out oxidative phosphorylation. In drosophila, FXN homolog (dfh) was identified, a much smaller genomic region than the human gene [57]. Its genomic composition just includes a single intron dividing the coding region into two exons. Moderate systemic reduction of *dfh* expression with RNA interference (RNAi) to levels of 30% of normal dfh-mRNA represents the situation in patients of FRDA, in which flies develop with normal embryonic stage, but shortened mean and maximum life span of 60 and 32%, respectively [58]. Reduction of FXN drastically compromises aconitase activity and increases the levels of reactive oxygen species (ROS) through inactivation of Fe–S enzymes of the respiratory chain [58]. Interestingly, the ectopic expression of enzymes that scavenge H_2O_2 suppresses the deleterious phenotypes associated with FXN deficiency [59]. In contrast, genetic augmentation with enzymes that scavenge superoxide is ineffective, suggesting that H_2O_2 is an important pathogenic substrate in FRDA with promising therapeutic implications (reviewed in [60]).

The genome of the nematode *Caenorhabditis elegans* is unusual among eukaryotes due to the operon organization, in which the *C. elegans FXN* gene, *frh*-1, is encoded in the operon CEOP2232 with complex structure and regulation [61]. Some work was carried out with *C. elegans* as a model for FRDA study. A transient knockdown by RNAi showed a consistent pleiotropic phenotype that includes poor growth, short lifespan, and increased sensitivity to oxidative stress [62]. Flavin adenine dinucleotide may rescue the phenotype of Frh-1 deficiency, very likely by enhancing the respiratory capacity [63]. Interestingly, the extended lifespan has been observed when *frh*-1 was knocking down by RNAi [64–66]. The difference may result from the efficiency of knockdown, i.e., the varied protein levels of frh-1 or the complexity of *frh*-1-containing operon.

Generation of a representative mouse model of FRDA is considered very important for the understanding of disease pathology and the testing of potential therapeutic strategies. Initially, a homozygous knockout of the murine *FXN* gene leads to early embryonic lethality without evidence of iron accumulation, indicating that FXN plays an important role during early embryonic development [55]. Later, a striated muscle Fxn-deficient line and a neuron/cardiac muscle FXN-deficient line were generated [67] to confirm the function of FXN involving in Fe–S cluster biogenesis. Though the ablation of Fxn in adult mice led to progressive neurological symptoms, the clinical symptoms could not well resemble FRDA. Since GAA repeat expansion within the first intron of the human *FXN* gene is the genetic basis of the disease, a mouse model harboring GAA repeats was created. The semizygous mice (knockin-knockout, KIKO) expressing 25–36% of wild-type FXN levels carry a (GAA)230 repeat expansion in the first intron of the mouse *Fxn* gene [68]. Unexpectedly, the mice did not present phenotypes of FRDA notwithstanding meiotically and mitotically stable GAA repeats [68]. Probably, the protein levels of Fxn are still above the threshold of the needs. In an attempt to generate a more relevant mouse model, human *FXN* YAC transgenic mice containing GAA repeat expansion derived from FRDA patient were created [69, 70]. These mice closely represent the first GAA repeat expansion-based FRDA models, which exhibit progressive FRDA-like pathology due to deficit of the important roles of FXN in mitochondrial function and iron metabolism. More recently, an inducible and reversible mouse model of Fxn deficiency was developed [71], which provides a broad use in therapeutic development and in refining our understanding of the relative contribution of reversible cellular dysfunction at different stages in disease.

7.4 FRDA and Iron

Iron relation was first evidenced for myocardial iron deposits in FRDA hearts [72]. Much later, iron involvement was recognized by the finding of deficiencies of aconitase and succinate dehydrogenase [28], both of which are mitochondrial iron–sulfur cluster-containing enzymes.

FXN deficiency leads to mitochondrial iron overload, defective energy supply, and generation of reactive oxygen species [67] (also see reviews [34, 73]). To date, no data provide a clear answer on the nature of the iron that is accumulated in affected mitochondria, but few interesting hints are available. The deposited mitochondrial iron was identified to be a biomineral aggregate distinctly different from mammalian ferritin [74] due to low expression of mitochondrial ferritin and high expression of mitochondrial importer mitoferrin 2 (MFRN2) [75]. What drives high expression of mitoferrin and how the excess toxic iron is trapped in the mitochondrial matrix are still unknown. It seems that mitochondrial iron overload is commonly taken place when Fe–S cluster biogenesis or transfer is disturbed. A similar increase of MFRN2 mRNA has been revealed in skeletal muscle biopsies from ISCU myopathy patients [76]. These results suggest the existence of a Fe–S cluster-dependent regulation of mitochondrial iron homeostasis. Does it mean that a particular transcription factor, likely, regulated by a Fe–S-dependent peptide or protein, modulates mitochondrial iron acquisition and retention [77]?

7.4.1 Iron in Nervous System

The neurological phenotypes of FRDA reflect lesions in the central nervous system and peripheral nervous system. The major lesions are located in the dorsal root ganglia of spinal cord and the dentate nucleus of the cerebellum [78]. The sensory peripheral nerves are often affected. Optic atrophy or hearing loss is relatively uncommon though it may be severe enough to cause deafness in approximately 1% of patients [79, 80]. Despite normal pure tone audiometry, patients with FRDA may have problems with speech perception [81]. Because dentate nucleus (DN) and dorsal root ganglions (DRGs) of individuals with FRDA are severely affected tissues, they have been extensively studied to investigate iron dysregulation by Dr. Koeppen AH [74]. Dentate nucleus is an iron-rich cerebellar structure that shows signs of neurodegeneration in patients with FRDA [78, 82, 83]. The much smaller size of the dentate nucleus in FRDA than in normal specimen is particularly apparent on a macro-stain for iron [84]. It lacks the distinct gray matter ribbon of the normal nucleus, and the intensity of the iron staining is lower than normal [82]. Further studies revealed that ferritin detected by immunohistochemistry is most abundant in oligodendroglia of the DN hilus and hypertrophic microglia in the atrophic DN gray matter of FRDA [85]. So far, the presence of mitochondrial iron accumulation in FRDA-affected neurons is unclear although it is common in heart of model organisms and FRDA patients.

Despite a report of difference in the MRI signals that suggests an overall increase of iron in the dentate nucleus of individuals with FRDA [86, 87], no difference in iron concentrations was demonstrated using autopsies [82]. However, modification of the expression of iron-related proteins such as transferrin receptor1, ferritins, and ferroportin was observed, thereby suggesting a change in iron metabolism [82], supporting that iron was relocating, likely, from dying neurons to microglia of dentate nucleus.

Similarly, DRGs from individuals with FRDA do not show overall iron concentrations above normal [88]. However, the expression of ferritin, the iron-storage protein, increases in satellite cells surrounding affected DRG neurons and decreases in DRG neurons [78, 88]. Meanwhile, not only the overall size reduction of subcapsular nerve cells in the DRG, but also the more proliferated glial cells surrounding the degenerated neurons are observed in DRG, in consistence with the much thicker satellite layer around neurons in patients than in normal subjects, whereas sural nerves in FRDA showed no convincing change in ferritin and ferroportin, militating against local iron dysmetabolism. Surprisingly, mitochondrial iron deposits have never been reported in neurons from FRDA individuals [89] and were not observed in an inducible conditional mouse model reproducing the neuronal phenotypes (PrPmice) [90]. The results stand out in contrast to the previously reported changes in dorsal spinal roots of patients with FRDA. Therefore, we are not sure about any indication on the primary involvement of iron dysregulation in the development of dorsal root ganglion in FRDA patients.

7.4.2 *Iron in Heart*

FRDA is an autosomal recessive disease with a complex neurological phenotype, but the most common cause of death is heart failure. Most FRDA patients succumb to cardiomyopathy, which progressively induces heart failure during the course of their disease. Heart is the first place to be revealed that iron accumulation occurs in FRDA patient tissues [72, 91]. It is relatively easy to find that the sections of the autopsy specimens display iron-positive granules in cardiac muscle fibers and all examined subjects presented Fe-positive inclusions though the degree of fiber hypertrophy and fibrosis varied [91]. More evidences showed that iron metabolism dysregulation occurred in heart autopsies of individuals with FRDA and the iron deposit is within mitochondria of cardiomyocytes [92]. However, the total iron obtained by assay of bulk digests of normal and FRDA left ventricular wall, right ventricular wall, and ventricular septum did not reveal a net increase [93] though a trend toward higher Fe concentrations was evident [91]. The mitochondrial iron accumulation and deposits were observed in the conditional mouse model reproducing the cardiac phenotype (MCK mouse) [67], which was thought to reflect progression.

Inflammation is an important factor in the pathogenesis of FRDA cardiomyopathy but may be more evident in advanced stages of the disease [91]. In comparison with the normal section of heart, the inflammatory nodules with macrophages in pathological areas contained granular iron and iron-laden phagocytes invade and replace sarcoplasm and myofibrils of single fibers. The further molecular evidence was that CD68-positive inflammatory cells express cytosolic ferritin and hepcidin, an iron-regulatory hormone [91]. The retention of iron in macrophages is a likely contributor to heart failure in FRDA due to the highly presented hepcidin to keep the iron from export by ferroportin. The unevenly distributed iron in heart suggested that iron accumulation progressed from a few small granules to coarse aggregates in phagocytized cardiomyocytes or infiltrated macrophages to result in elevated levels of iron in highly localized, randomly distributed regions of the heart.

Altogether, these observations led to the early assumption of a pathophysiological implication of iron-dependent pathways in FRDA. However, it is not clear why the restricted accumulation of cardiac iron in FRDA should be much more damaging than global iron overload and what drives the heterogeneous distribution of iron in heart with variance from systemic iron overload.

7.4.3 Iron in Pancreas

Cardiomyopathy or neurological signs are the first clinical manifestation, whereas the late onset of diabetes mellitus is often observed in the course of the illness, but not present in all patients. Early occurrence of diabetes and rapid progression of ataxia in FRDA patient were also observed [94]. The incidence rates vary between 8 and 32% indicating that FXN deficiency is a risk factor of diabetes. In a recent study of Friedreich ataxia, a much higher percentage (49%) of the participants had impaired fasting glucose and/or impaired glucose tolerance without a preexisting clinical diagnosis of diabetes [95]. Interestingly, in non-FRDA subjects, the level of FXN expression is also much lower in T2DM islets [96]. In the same study, a positive correlation was apparent between response to glucose stimulation and response to FXN gene expression, suggesting that FXN levels of human pancreatic islets contribute to the prevention from T2DM. And a shared mechanism of apoptosis between β-cells and neurons in FRDA is revealed [97] that FXN deficiency in β-cells induces mitochondrial oxidative stress and activates the intrinsic pathway of apoptosis. Puzzlingly, no mitochondrial iron accumulation has been reported in β-cells of pancreatic islets in FRDA patients.

Why FXN deficiency mainly compromises neurons, cardiomyocytes, and pancreatic β-cells is elusive. Although these cells have a distinct embryonic origin, they need mitochondria to fulfill many functional activities. In addition, three types of cells in those tissues are very long lived. What very fundamental features are common in the long-lived cells may tell us more the pathogenesis of FRDA.

7.5 FRDA Treatment by Manipulation of Iron Status

At present, there is no any approved cure treatment for this progressive condition. However, the efforts for a potential pharmacological treatment have been made considerably in the past two decades with many potential therapeutic agents proposed to delay disease progression (reviewed in Ref. [98]). An optimal therapy in FRDA would consist of providing sufficient intracellular FXN to restore the physiological functions of cells, including the gene therapy [99, 100] and drugs increasing FXN expression [101]. Interestingly, two very recent studies have reported that bone marrow transplantation stimulates neural repair and ameliorates deficits in FRDA mice [102, 103]. Here, we only take attention on iron-related therapies in mouse model and human.

It is often observed that iron accumulation in mitochondria occurs in case of impairment of Fe–S cluster biogenesis (reviewed in [29]). Iron accumulation was first observed in the heart of FRDA patients [72] as a secondary effect of FXN deficiency [55, 67]. The excess iron may play destructive effects in several cellular functions due to its unique physiochemical properties. Interestingly, it seems still controversial in the central neuronal system [82, 87]. One possibility could be the ratio of Fe^{2+}/Fe^{3+} shifted as observed in Parkinson disease (PD) patients [104]. In this mentioned study, iron ratio was reported to be shifted from 1:2 in healthy subjects to 2:1 in PD patients in substantia nigra pars compacta regions of the brain. The total iron content measured with some common methods may not distinguish the Fe^{2+} and Fe^{3+}. The increase in the concentration of highly reactive iron species, Fe^{2+}, results in excess production of ROS via Fenton chemistry. Nevertheless, iron chelation therapy was proposed and assessed by Cabantchik group [86]. In that study, they found that a membrane-permeant and orally administered chelator, deferiprone (DFP), reduced neuropathy and ataxic gait only in the youngest FRDA patients by shuttling chelated iron from cells to transferrin to reduce brain iron accumulation without systemic anemia. The results indicate that DFP is able to cross the blood–brain barrier. It was commonly observed in a few clinical trials that DFP administration at low dose was associated with a significant reduction in cardiac hypertrophy, or an abnormal enlargement, or thickening, of the heart muscle (reviewed in [105]). The effects of long-term treatment at low doses are suggested to be worthy to be evaluated later on since most of the patients die of cardiac failure. However, the results were somehow puzzling as different doses showed opposite effects, if any (reviewed in [106]). Moreover, our previous results demonstrated that iron deficiency downregulated *FXN* expression [107] and reduced aconitase activity even with low concentration of DFP (Li, unpublished data) in cells derived from healthy controls and FRDA patients. Therefore, the rationale is further questioned behind the use of chelation therapy in FRDA. Larger studies with hard endpoints are still missing.

Things might get a turn for a hope. A combined therapy of a low dose of DFP with idebenone is relatively safe and might improve neurological function and heart hypertrophy in a 6-month randomized, double-blind placebo-controlled study [108]. Some novel bifunctional or multifunctional agents have been synthesized with the ability of iron chelation and monoamine oxidase inhibitory (MAOI) activities for ironing out iron from the brain and preventing iron-induced Fenton reaction from hydroxyl radical toxicity [109]. Though these pharmacophores as iron chelators are being examined for treatment in other neurodegenerative diseases with regional increase of iron levels (reviewed in [110, 111]), such as Parkinson and Alzheimer diseases, the similar strategies are worth to be tried in the treatment of FRDA. Besides iron chelation, the radical scavenging ability, the mitochondrial activity enhancing, and upregulation of FXN expression may also be under consideration to be combined. Recently, we used a mitochondria-targeted agent SS-31, a tetrapeptide with the ability to enrich in mitochondria more than 1000-folds than in other organelles within cells by binding cardiolipin [112]. This binding is thought to maintain the mitochondrial structure and integrity to improve the mitochondrial function. Interestingly, not only as observed to reduce the oxidative stress, but also increase of FXN expression at translational levels was revealed in a model of cells derived from FRDA patients [101]. When we used the GAA-containing mouse model [69], the FRDA phenotypes were reversed and the similar mechanism was revealed to that observed in cell model of FRDA (in preparation for publication). Therefore, drugs designed based on iron chelation, even on more accurate characteristics for iron shuttling between organelles, are not enough. Due to the important role of FXN in mitochondrial function, development of therapeutic strategies should end at targeting more pathways in combination with a few drugs or simply end at elevating FXN protein levels. Gene therapy and transplantation of stem cells are also recommended.

7.6 Challenges Associated with Iron Chelators

Iron involvement in the pathophysiology of FRDA does not seem to be a simple event of iron toxicity. The consequence of neurodegeneration is solidly connected with mitochondrial dysfunction, rather than with iron accumulation in the disease. Particularly, the iron accumulation only occurs in some tissues, more specifically in mitochondria of the affected cells in these tissues. Though DFP may shuttle iron between subcellular compartments or transfer iron from iron-overloaded cells to extracellular apotransferrin and pre-erythroid cells for heme synthesis [106], DFP still is a systemic iron chelator, with which iron is complexed and excreted in the urine within 5–6 h of administration (https://pubchem.ncbi.nlm.nih.gov). It is speculated that the iron stores in the body would decrease when DFP was administrated though the anemia was not clearly observed in a 6-months trial [86]. We observed that DFP treatment reduced the mitochondrial aconitase activity even in a low dose in both wild-type and mutant cells derived from FRDA patients (unpublished). On the other hand, excess mitochondrial iron uptake and sequestration cause cytosolic iron depletion, which can further diminish FXN expression by decreasing FXN transcription [107]. Thus, the decrease in FXN levels caused by the trinucleotide repeat may be worsened by DFP administration. As cells accumulate mitochondrial iron and deplete cytosolic iron stores, we postulate that it may form a negative feedback loop to contribute to progression of disease in long-lived cells, particularly neurons and cardiomyocytes [107]. On the contrary, a seemingly paradox method was suggested that addition of an iron supplementation regime that would replenish cytosolic iron might be of value in conjunction with iron chelators, such as DFP that would reduce mitochondrial iron stores [107], or with antioxidants that would mitigate the effects of mitochondrial iron overload [113]. These therapeutic approaches could potentially increase FXN expression or at least maintain FXN levels, in affected cells in FRDA patients, which management might improve the quality of life and slow the disease progression.

7.7 Summary

The Friedreich's Ataxia Research Alliance (FARA) provides a comprehensive outline of current therapeutic agents with a treatment pipeline (http://www.curefa.org/pipeline). No study has successfully achieved well-accepted endpoint thus far because the exact function of FXN is still undermining. Though it is well accepted that the primary consequence of FXN deficiency is mitochondrial dysfunction, any drugs that can promote mitochondrial function do not widely target on all affected steps. The combination of a few mitochondrial function-improving drugs might be an option to be considered. Meanwhile, the defect of Fe–S biogenesis compromises a large range of enzymes within and extra-mitochondria; therefore, drugs targeting extra-mitochondrial process may be also a wise choice.

References

1. Colombo R, Carobene A (2000) Age of the intronic GAA triplet repeat expansion mutation in Friedreich ataxia. Hum Genet 106:455–458
2. Nieto A, Hernández-Torres A, Pérez-Flores J, Montón F (2018) Depressive symptoms in Friedreich ataxia. Int J Clin Health Psychol 18:18–26
3. Campuzano V, Montermini L, Molto MD, Pianese L, Cossee M, Cavalcanti F, Monros E, Rodius F, Duclos F, Monticelli A, Zara F, Canizares J, Koutnikova H, Bidichandani SI, Gellera C, Brice A, Trouillas P, De Michele G, Filla A, De Frutos R, Palau F, Patel PI, Di Donato S, Mandel JL, Cocozza S, Koenig M, Pandolfo M (1996) Friedreich's ataxia: autosomal recessive disease caused by an intronic GAA triplet repeat expansion. Science 271:1423–1427
4. Cossee M, Durr A, Schmitt M, Dahl N, Trouillas P, Allinson P, Kostrzewa M, Nivelon-Chevallier A, Gustavson KH, Kohlschutter A, Muller U, Mandel JL, Brice A, Koenig M, Cavalcanti F, Tammaro A, De Michele G, Filla A, Cocozza S, Labuda M, Montermini L, Poirier J, Pandolfo M (1999) Friedreich's ataxia: point mutations and clinical presentation of compound heterozygotes. Ann Neurol 45:200–206
5. De Castro M, Garcia-Planells J, Monros E, Canizares J, Vazquez-Manrique R, Vilchez JJ, Urtasun M, Lucas M, Navarro G, Izquierdo G, Molto MD, Palau F (2000) Genotype and phenotype analysis of Friedreich's ataxia compound heterozygous patients. Hum Genet 106:86–92

6. Anheim M, Mariani LL, Calvas P, Cheuret E, Zagnoli F, Odent S, Seguela C, Marelli C, Fritsch M, Delaunoy JP, Brice A, Durr A, Koenig M (2012) Exonic deletions of FXN and early-onset Friedreich ataxia. Arch Neurol 69:912–916
7. Barcia G, Rachid M, Magen M, Assouline Z, Koenig M, Funalot B, Barnerias C, Rotig A, Munnich A, Bonnefont JP, Steffann J (2018) Pitfalls in molecular diagnosis of Friedreich ataxia. Eur J Med Genet
8. Gottesfeld JM (2007) Small molecules affecting transcription in Friedreich ataxia. Pharmacol Ther 116:236–248
9. Schmucker S, Puccio H (2010) Understanding the molecular mechanisms of Friedreich's ataxia to develop therapeutic approaches. Hum Mol Genet 19:R103–R110
10. Al-Mahdawi S, Pinto RM, Ismail O, Varshney D, Lymperi S, Sandi C, Trabzuni D, Pook M (2008) The Friedreich ataxia GAA repeat expansion mutation induces comparable epigenetic changes in human and transgenic mouse brain and heart tissues. Hum Mol Genet 17:735–746
11. Castaldo I, Pinelli M, Monticelli A, Acquaviva F, Giacchetti M, Filla A, Sacchetti S, Keller S, Avvedimento VE, Chiariotti L, Cocozza S (2008) DNA methylation in intron 1 of the frataxin gene is related to GAA repeat length and age of onset in Friedreich ataxia patients. J Med Genet 45:808–812
12. Greene E, Mahishi L, Entezam A, Kumari D, Usdin K (2007) Repeat-induced epigenetic changes in intron 1 of the frataxin gene and its consequences in Friedreich ataxia. Nucleic Acids Res 35:3383–3390
13. Rai M, Soragni E, Jenssen K, Burnett R, Herman D, Coppola G, Geschwind DH, Gottesfeld JM, Pandolfo M (2008) HDAC inhibitors correct frataxin deficiency in a Friedreich ataxia mouse model. PLoS ONE 3:e1958
14. Soragni E, Herman D, Dent SY, Gottesfeld JM, Wells RD, Napierala M (2008) Long intronic GAA*TTC repeats induce epigenetic changes and reporter gene silencing in a molecular model of Friedreich ataxia. Nucleic Acids Res 36:6056–6065
15. Li K, Singh A, Crooks DR, Dai X, Cong Z, Pan L, Ha D, Rouault TA (2010) Expression of human frataxin is regulated by transcription factors SRF and TFAP2. PLoS ONE 5:e12286
16. Guccini I, Serio D, Condo I, Rufini A, Tomassini B, Mangiola A, Maira G, Anile C, Fina D, Pallone F, Mongiardi MP, Levi A, Ventura N, Testi R, Malisan F (2011) Frataxin participates to the hypoxia-induced response in tumors. Cell Death Dis 2:e123
17. Shimizu R, Lan NN, Tai TT, Adachi Y, Kawazoe A, Mu A, Taketani S (2014) p53 directly regulates the transcription of the human frataxin gene and its lack of regulation in tumor cells decreases the utilization of mitochondrial iron. Gene 551:79–85
18. Condo I, Ventura N, Malisan F, Rufini A, Tomassini B, Testi R (2007) In vivo maturation of human frataxin. Hum Mol Genet 16:1534–1540
19. Acquaviva F, De Biase I, Nezi L, Ruggiero G, Tatangelo F, Pisano C, Monticelli A, Garbi C, Acquaviva AM, Cocozza S (2005) Extra-mitochondrial localisation of frataxin and its association with IscU1 during enterocyte-like differentiation of the human colon adenocarcinoma cell line Caco-2. J Cell Sci 118:3917–3924
20. Condo I, Malisan F, Guccini I, Serio D, Rufini A, Testi R (2010) Molecular control of the cytosolic aconitase/IRP1 switch by extramitochondrial frataxin. Hum Mol Genet 19:1221–1229
21. Guo L, Wang Q, Weng L, Hauser LA, Strawser CJ, Mesaros C, Lynch DR, Blair IA (2018) Characterization of a new N-terminally acetylated extra-mitochondrial isoform of frataxin in human erythrocytes. Sci Rep 8:17043
22. Xia H, Cao Y, Dai X, Marelja Z, Zhou D, Mo R, Al-Mahdawi S, Pook MA, Leimkuhler S, Rouault TA, Li K (2012) Novel frataxin isoforms may contribute to the pathological mechanism of Friedreich ataxia. PLoS ONE 7:e47847
23. Schmucker S, Martelli A, Colin F, Page A, Wattenhofer-Donze M, Reutenauer L, Puccio H (2011) Mammalian frataxin: an essential function for cellular viability through an interaction with a preformed ISCU/NFS1/ISD11 iron-sulfur assembly complex. PLoS ONE 6:e16199
24. Perez-Luz S, Gimenez-Cassina A, Fernandez-Frias I, Wade-Martins R, Diaz-Nido J (2015) Delivery of the 135 kb human frataxin genomic DNA locus gives rise to different frataxin isoforms. Genomics 106:176–182

25. Gakh O, Bedekovics T, Duncan SF, Smith DY, Berkholz DS, Isaya G (2010) Normal and Friedreich ataxia cells express different isoforms of frataxin with complementary roles in iron-sulfur cluster assembly. J Biol Chem 285:38486–38501
26. Musco G, Stier G, Kolmerer B, Adinolfi S, Martin S, Frenkiel T, Gibson T, Pastore A (2000) Towards a structural understanding of Friedreich's ataxia: the solution structure of frataxin. Structure 8:695–707
27. Dhe-Paganon S, Shigeta R, Chi YI, Ristow M, Shoelson SE (2000) Crystal structure of human frataxin. J Biol Chem 275:30753–30756
28. Rotig A, de Lonlay P, Chretien D, Foury F, Koenig M, Sidi D, Munnich A, Rustin P (1997) Aconitase and mitochondrial iron-sulphur protein deficiency in Friedreich ataxia. Nat Genet 17:215–217
29. Stehling O, Lill R (2013) The role of mitochondria in cellular iron-sulfur protein biogenesis: mechanisms, connected processes, and diseases. Cold Spring Harb Perspect Biol 5:a011312
30. Gerber J, Muhlenhoff U, Lill R (2003) An interaction between frataxin and Isu1/Nfs1 that is crucial for Fe/S cluster synthesis on Isu1. EMBO Rep 4:906–911
31. Layer G, Ollagnier-de Choudens S, Sanakis Y, Fontecave M (2006) Iron-sulfur cluster biosynthesis: characterization of Escherichia coli CYaY as an iron donor for the assembly of [2Fe–2S] clusters in the scaffold IscU. J Biol Chem 281:16256–16263
32. Shan Y, Napoli E, Cortopassi G (2007) Mitochondrial frataxin interacts with ISD11 of the NFS1/ISCU complex and multiple mitochondrial chaperones. Hum Mol Genet 16:929–941
33. Li H, Gakh O, Smith DY, Isaya G (2009) Oligomeric yeast frataxin drives assembly of core machinery for mitochondrial iron-sulfur cluster synthesis. J Biol Chem 284:21971–21980
34. Alsina D, Purroy R, Ros J, Tamarit J (2018) Iron in Friedreich Ataxia: a central role in the pathophysiology or an epiphenomenon? Pharmaceuticals (Basel) 11
35. Gakh O, Ranatunga W, Smith DY, Ahlgren EC, Al-Karadaghi S, Thompson JR, Isaya G (2016) Architecture of the human mitochondrial iron-sulfur cluster assembly machinery. J Biol Chem 291:21296–21321
36. Armas AM, Balparda M, Terenzi A, Busi MV, Pagani MA, Gomez-Casati DF (2019) Ferrochelatase activity of plant frataxin. Biochimie 156:118–122
37. Cook JD, Bencze KZ, Jankovic AD, Crater AK, Busch CN, Bradley PB, Stemmler AJ, Spaller MR, Stemmler TL (2006) Monomeric yeast frataxin is an iron-binding protein. Biochemistry 45:7767–7777
38. Yoon T, Cowan JA (2003) Iron-sulfur cluster biosynthesis. Characterization of frataxin as an iron donor for assembly of [2Fe–2S] clusters in ISU-type proteins. J Am Chem Soc 125:6078–6084
39. Bencze KZ, Yoon T, Millan-Pacheco C, Bradley PB, Pastor N, Cowan JA, Stemmler TL (2007) Human frataxin: iron and ferrochelatase binding surface. Chem Commun (Camb) 1798–1800
40. Yoon T, Cowan JA (2004) Frataxin-mediated iron delivery to ferrochelatase in the final step of heme biosynthesis. J Biol Chem 279:25943–25946
41. Steinkellner H, Singh HN, Muckenthaler MU, Goldenberg H, Moganty RR, Scheiber-Mojdehkar B, Sturm B (2017) No changes in heme synthesis in human Friedreich s ataxia erythroid progenitor cells. Gene 621:5–11
42. Adamec J, Rusnak F, Owen WG, Naylor S, Benson LM, Gacy AM, Isaya G (2000) Iron-dependent self-assembly of recombinant yeast frataxin: implications for Friedreich ataxia. Am J Hum Genet 67:549–562
43. Ahlgren EC, Fekry M, Wiemann M, Soderberg CA, Bernfur K, Gakh O, Rasmussen M, Hojrup P, Emanuelsson C, Isaya G, Al-Karadaghi S (2017) Iron-induced oligomerization of human FXN81-210 and bacterial CyaY frataxin and the effect of iron chelators. PLoS ONE 12:e0188937
44. Schagerlof U, Elmlund H, Gakh O, Nordlund G, Hebert H, Lindahl M, Isaya G, Al-Karadaghi S (2008) Structural basis of the iron storage function of frataxin from single-particle reconstruction of the iron-loaded oligomer. Biochemistry 47:4948–4954
45. Park S, Gakh O, Mooney SM, Isaya G (2002) The ferroxidase activity of yeast frataxin. J Biol Chem 277:38589–38595

46. Soderberg CA, Rajan S, Shkumatov AV, Gakh O, Schaefer S, Ahlgren EC, Svergun DI, Isaya G, Al-Karadaghi S (2013) The molecular basis of iron-induced oligomerization of frataxin and the role of the ferroxidation reaction in oligomerization. J Biol Chem 288:8156–8167
47. Sutak R, Seguin A, Garcia-Serres R, Oddou JL, Dancis A, Tachezy J, Latour JM, Camadro JM, Lesuisse E (2012) Human mitochondrial ferritin improves respiratory function in yeast mutants deficient in iron-sulfur cluster biogenesis, but is not a functional homologue of yeast frataxin. Microbiologyopen 1:95–104
48. Vaubel RA, Isaya G (2013) Iron-sulfur cluster synthesis, iron homeostasis and oxidative stress in Friedreich ataxia. Mol Cell Neurosci 55:50–61
49. Shaw GC, Cope JJ, Li L, Corson K, Hersey C, Ackermann GE, Gwynn B, Lambert AJ, Wingert RA, Traver D, Trede NS, Barut BA, Zhou Y, Minet E, Donovan A, Brownlie A, Balzan R, Weiss MJ, Peters LL, Kaplan J, Zon LI, Paw BH (2006) Mitoferrin is essential for erythroid iron assimilation. Nature 440:96–100
50. Tsai CL, Barondeau DP (2010) Human frataxin is an allosteric switch that activates the Fe–S cluster biosynthetic complex. Biochemistry 49:9132–9139
51. Colin F, Martelli A, Clemancey M, Latour JM, Gambarelli S, Zeppieri L, Birck C, Page A, Puccio H, Ollagnier de Choudens S (2015) Mammalian frataxin controls sulfur production and iron entry during de novo Fe4S4 cluster assembly. J Am Chem Soc 135:733–740
52. Bridwell-Rabb J, Fox NG, Tsai CL, Winn AM, Barondeau DP (2014) Human frataxin activates Fe–S cluster biosynthesis by facilitating sulfur transfer chemistry. Biochemistry 53:4904–4913
53. Foury F, Cazzalini O (1997) Deletion of the yeast homologue of the human gene associated with Friedreich's ataxia elicits iron accumulation in mitochondria. FEBS Lett 411:373–377
54. Llorens JV, Soriano S, Calap-Quintana P, Gonzalez-Cabo P, Molto MD (2019) The role of Iron in Friedreich's ataxia: insights from studies in human tissues and cellular and animal models. Front Neurosci 13:75
55. Cossee M, Puccio H, Gansmuller A, Koutnikova H, Dierich A, LeMeur M, Fischbeck K, Dolle P, Koenig M (2000) Inactivation of the Friedreich ataxia mouse gene leads to early embryonic lethality without iron accumulation. Hum Mol Genet 9:1219–1226
56. Babcock M, de Silva D, Oaks R, Davis-Kaplan S, Jiralerspong S, Montermini L, Pandolfo M, Kaplan J (1997) Regulation of mitochondrial iron accumulation by Yfh1p, a putative homolog of frataxin. Science 276:1709–1712
57. Canizares J, Blanca JM, Navarro JA, Monros E, Palau F, Molto MD (2000) dfh is a Drosophila homolog of the Friedreich's ataxia disease gene. Gene 256:35–42
58. Llorens JV, Navarro JA, Martinez-Sebastian MJ, Baylies MK, Schneuwly S, Botella JA, Molto MD (2007) Causative role of oxidative stress in a Drosophila model of Friedreich ataxia. FASEB J. 21:333–344
59. Anderson PR, Kirby K, Orr WC, Hilliker AJ, Phillips JP (2008) Hydrogen peroxide scavenging rescues frataxin deficiency in a Drosophila model of Friedreich's ataxia. Proc Natl Acad Sci U S A 105:611–616
60. Calap-Quintana P, Navarro JA, Gonzalez-Fernandez J, Martinez-Sebastian MJ, Molto MD, Llorens JV (2018) Drosophila melanogaster Models of Friedreich's Ataxia. Biomed Res Int 2018:5065190
61. Vazquez-Manrique RP, Gonzalez-Cabo P, Ortiz-Martin I, Ros S, Baylis HA, Palau F (2007) The frataxin-encoding operon of *Caenorhabditis elegans* shows complex structure and regulation. Genomics 89:392–401
62. Vazquez-Manrique RP, Gonzalez-Cabo P, Ros S, Aziz H, Baylis HA, Palau F (2006) Reduction of *caenorhabditis elegans* frataxin increases sensitivity to oxidative stress, reduces lifespan, and causes lethality in a mitochondrial complex II mutant. FASEB J. 20:172–174
63. Gonzalez-Cabo P, Ros S, Palau F (2010) Flavin adenine dinucleotide rescues the phenotype of frataxin deficiency. PLoS ONE 5:e8872
64. Ventura N, Rea S, Henderson ST, Condo I, Johnson TE, Testi R (2005) Reduced expression of frataxin extends the lifespan of *Caenorhabditis elegans*. Aging Cell 4:109–112

65. Schiavi A, Torgovnick A, Kell A, Megalou E, Castelein N, Guccini I, Marzocchella L, Gelino S, Hansen M, Malisan F, Condo I, Bei R, Rea SL, Braeckman BP, Tavernarakis N, Testi R, Ventura N (2013) Autophagy induction extends lifespan and reduces lipid content in response to Frataxin silencing in *C. elegans*. Exp Gerontol 48:191–201
66. Schiavi A, Maglioni S, Palikaras K, Shaik A, Strappazzon F, Brinkmann V, Torgovnick A, Castelein N, De Henau S, Braeckman BP, Cecconi F, Tavernarakis N, Ventura N (2015) Iron-starvation-induced mitophagy mediates lifespan extension upon mitochondrial stress in *C. elegans*. Curr Biol 25:1810–1822
67. Puccio H, Simon D, Cossee M, Criqui-Filipe P, Tiziano F, Melki J, Hindelang C, Matyas R, Rustin P, Koenig M (2001) Mouse models for Friedreich ataxia exhibit cardiomyopathy, sensory nerve defect and Fe–S enzyme deficiency followed by intramitochondrial iron deposits. Nat Genet 27:181–186
68. Miranda CJ, Santos MM, Ohshima K, Smith J, Li L, Bunting M, Cossee M, Koenig M, Sequeiros J, Kaplan J, Pandolfo M (2002) Frataxin knockin mouse. FEBS Lett 512:291–297
69. Anjomani Virmouni S, Ezzatizadeh V, Sandi C, Sandi M, Al-Mahdawi S, Chutake Y, Pook MA (2015) A novel GAA-repeat-expansion-based mouse model of Friedreich's ataxia. Dis Model Mech 8:225–235
70. Al-Mahdawi S, Pinto RM, Ruddle P, Carroll C, Webster Z, Pook M (2004) GAA repeat instability in Friedreich ataxia YAC transgenic mice. Genomics 84:301–310
71. Chandran V, Gao K, Swarup V, Versano R, Dong H, Jordan MC, Geschwind DH (2017) Inducible and reversible phenotypes in a novel mouse model of Friedreich's Ataxia. Elife 6
72. Sanchez-Casis G, Cote M, Barbeau A (1976) Pathology of the heart in Friedreich's ataxia: review of the literature and report of one case. Can J Neurol Sci Le J Can des Sci Neurologiques 3:349–354
73. Burk K (2017) Friedreich Ataxia: current status and future prospects. Cerebellum Ataxias 4:4
74. Whitnall M, Suryo Rahmanto Y, Huang ML, Saletta F, Lok HC, Gutierrez L, Lazaro FJ, Fleming AJ, St Pierre TG, Mikhael MR, Ponka P, Richardson DR (2012) Identification of nonferritin mitochondrial iron deposits in a mouse model of Friedreich ataxia. Proc Natl Acad Sci U S A 109:20590–20595
75. Huang ML, Becker EM, Whitnall M, Suryo Rahmanto Y, Ponka P, Richardson DR (2009) Elucidation of the mechanism of mitochondrial iron loading in Friedreich's ataxia by analysis of a mouse mutant. Proc Natl Acad Sci U S A 106:16381–16386
76. Crooks DR, Natarajan TG, Jeong SY, Chen C, Park SY, Huang H, Ghosh MC, Tong WH, Haller RG, Wu C, Rouault TA (2014) Elevated FGF21 secretion, PGC-1alpha and ketogenic enzyme expression are hallmarks of iron-sulfur cluster depletion in human skeletal muscle. Hum Mol Genet 23:24–39
77. Rouault TA (2016) Mitochondrial iron overload: causes and consequences. Curr Opin Genet Dev 38:31–37
78. Koeppen AH, Mazurkiewicz JE (2013) Friedreich ataxia: neuropathology revised. J Neuropathol Exp Neurol 72:78–90
79. Harding AE (1981) Friedreich's ataxia: a clinical and genetic study of 90 families with an analysis of early diagnostic criteria and intrafamilial clustering of clinical features. Brain 104:589–620
80. Rance G, Corben L, Delatycki M (2012) Auditory processing deficits in children with Friedreich ataxia. J Child Neurol 27:1197–1203
81. Rance G, Fava R, Baldock H, Chong A, Barker E, Corben L, Delatycki MB (2008) Speech perception ability in individuals with Friedreich ataxia. Brain 131:2002–2012
82. Koeppen AH, Michael SC, Knutson MD, Haile DJ, Qian J, Levi S, Santambrogio P, Garrick MD, Lamarche JB (2007) The dentate nucleus in Friedreich's ataxia: the role of iron-responsive proteins. Acta Neuropathol 114:163–173
83. Ward PGD, Harding IH, Close TG, Corben LA, Delatycki MB, Storey E, Georgiou-Karistianis N, Egan GF (2019) Longitudinal evaluation of iron concentration and atrophy in the dentate nuclei in friedreich ataxia. Mov Disord 34:335–343

84. Koeppen AH (2011) Friedreich's ataxia: pathology, pathogenesis, and molecular genetics. J Neurol Sci 303:1–12
85. Koeppen AH, Ramirez RL, Yu D, Collins SE, Qian J, Parsons PJ, Yang KX, Chen Z, Mazurkiewicz JE, Feustel PJ (2012) Friedreich's ataxia causes redistribution of iron, copper, and zinc in the dentate nucleus. Cerebellum 11:845–860
86. Boddaert N, Le Quan Sang KH, Rotig A, Leroy-Willig A, Gallet S, Brunelle F, Sidi D, Thalabard JC, Munnich A, Cabantchik ZI (2007) Selective iron chelation in Friedreich ataxia: biologic and clinical implications. Blood 110:401–408
87. Waldvogel D, van Gelderen P, Hallett M (1999) Increased iron in the dentate nucleus of patients with Friedrich's ataxia. Ann Neurol 46:123–125
88. Koeppen AH, Morral JA, Davis AN, Qian J, Petrocine SV, Knutson MD, Gibson WM, Cusack MJ, Li D (2009) The dorsal root ganglion in Friedreich's ataxia. Acta Neuropathol 118:763–776
89. Morral JA, Davis AN, Qian J, Gelman BB, Koeppen AH (2010) Pathology and pathogenesis of sensory neuropathy in Friedreich's ataxia. Acta Neuropathol 120:97–108
90. Simon D, Seznec H, Gansmuller A, Carelle N, Weber P, Metzger D, Rustin P, Koenig M, Puccio H (2004) Friedreich ataxia mouse models with progressive cerebellar and sensory ataxia reveal autophagic neurodegeneration in dorsal root ganglia. J Neurosci 24:1987–1995
91. Koeppen AH, Ramirez RL, Becker AB, Bjork ST, Levi S, Santambrogio P, Parsons PJ, Kruger PC, Yang KX, Feustel PJ, Mazurkiewicz JE (2015) The pathogenesis of cardiomyopathy in Friedreich ataxia. PLoS ONE 10:e0116396
92. Michael S, Petrocine SV, Qian J, Lamarche JB, Knutson MD, Garrick MD, Koeppen AH (2006) Iron and iron-responsive proteins in the cardiomyopathy of Friedreich's ataxia. Cerebellum 5:257–267
93. Kruger PC, Yang KX, Parsons PJ, Becker AB, Feustel PJ, Koeppen AH (2016) Abundance and significance of Iron, Zinc, Copper, and Calcium in the hearts of patients with Friedreich Ataxia. Am J Cardiol 118:127–131
94. Garg M, Kulkarni SD, Shah KN, Hegde AU (2017) Diabetes Mellitus as the Presenting Feature of Friedreich's Ataxia. J Neurosci Rural Pract 8:S117–S119
95. Cnop M, Igoillo-Esteve M, Rai M, Begu A, Serroukh Y, Depondt C, Musuaya AE, Marhfour I, Ladriere L, Moles Lopez X, Lefkaditis D, Moore F, Brion JP, Cooper JM, Schapira AH, Clark A, Koeppen AH, Marchetti P, Pandolfo M, Eizirik DL, Fery F (2012) Central role and mechanisms of beta-cell dysfunction and death in friedreich ataxia-associated diabetes. Ann Neurol 72:971–982
96. Del Guerra S, D'Aleo V, Gualtierotti G, Pandolfi R, Boggi U, Vistoli F, Barnini S, Filipponi F, Del Prato S, Lupi R (2012) Evidence for a role of frataxin in pancreatic islets isolated from multi-organ donors with and without type 2 diabetes mellitus. Horm Metab Res 44:471–475
97. Igoillo-Esteve M, Gurgul-Convey E, Hu A, Romagueira Bichara Dos Santos L, Abdulkarim B, Chintawar S, Marselli L, Marchetti P, Jonas JC, Eizirik DL, Pandolfo M, Cnop M (2015) Unveiling a common mechanism of apoptosis in beta-cells and neurons in Friedreich's ataxia. Hum Mol Genet 24:2274–2286
98. Tai G, Corben LA, Yiu EM, Milne SC, Delatycki MB (2018) Progress in the treatment of Friedreich ataxia. Neurol Neurochir Pol 52:129–139
99. Perdomini M, Belbellaa B, Monassier L, Reutenauer L, Messaddeq N, Cartier N, Crystal RG, Aubourg P, Puccio H (2014) Prevention and reversal of severe mitochondrial cardiomyopathy by gene therapy in a mouse model of Friedreich's ataxia. Nat Med 20:542–547
100. Piguet F, de Montigny C, Vaucamps N, Reutenauer L, Eisenmann A, Puccio H (2018) Rapid and complete reversal of sensory Ataxia by gene therapy in a novel model of Friedreich Ataxia. Mol Ther 26:1940–1952
101. Zhao H, Li H, Hao S, Chen J, Wu J, Song C, Zhang M, Qiao T, Li K (2017) Peptide SS-31 upregulates frataxin expression and improves the quality of mitochondria: implications in the treatment of Friedreich ataxia. Sci Rep 7:9840
102. Rocca CJ, Goodman SM, Dulin JN, Haquang JH, Gertsman I, Blondelle J, Smith JLM, Heyser CJ, Cherqui S (2017) Transplantation of wild-type mouse hematopoietic stem and progenitor cells ameliorates deficits in a mouse model of Friedreich's ataxia. Sci Transl Med 9

103. Kemp KC, Hares K, Redondo J, Cook AJ, Haynes HR, Burton BR, Pook MA, Rice CM, Scolding NJ, Wilkins A (2018) Bone marrow transplantation stimulates neural repair in Friedreich's ataxia mice. Ann Neurol 83:779–793
104. Riederer P, Rausch WD, Schmidt B, Kruzik P, Konradi C, Sofic E, Danielczyk W, Fischer M, Ogris E (1988) Biochemical fundamentals of Parkinson's disease. Mt Sinai J Med 55:21–28
105. Nunez MT, Chana-Cuevas P (2018) New perspectives in iron chelation therapy for the treatment of neurodegenerative diseases. Pharmaceuticals (Basel) 11
106. Pandolfo M, Hausmann L (2013) Deferiprone for the treatment of Friedreich's ataxia. J Neurochem 126(Suppl 1):142–146
107. Li K, Besse EK, Ha D, Kovtunovych G, Rouault TA (2008) Iron-dependent regulation of frataxin expression: implications for treatment of Friedreich ataxia. Hum Mol Genet 17:2265–2273
108. Elincx-Benizri S, Glik A, Merkel D, Arad M, Freimark D, Kozlova E, Cabantchik I, Hassin-Baer S (2016) Clinical experience with deferiprone treatment for Friedreich Ataxia. J Child Neurol 31:1036–1040
109. Youdim MB, Fridkin M, Zheng H (2004) Novel bifunctional drugs targeting monoamine oxidase inhibition and iron chelation as an approach to neuroprotection in Parkinson's disease and other neurodegenerative diseases. J Neural Transm (Vienna) 111:1455–1471
110. Singh YP, Pandey A, Vishwakarma S, Modi G (2018) A review on iron chelators as potential therapeutic agents for the treatment of Alzheimer's and Parkinson's diseases. Mol Divers
111. Youdim MB, Fridkin M, Zheng H (2005) Bifunctional drug derivatives of MAO-B inhibitor rasagiline and iron chelator VK-28 as a more effective approach to treatment of brain ageing and ageing neurodegenerative diseases. Mech Ageing Dev 126:317–326
112. Szeto HH (2014) First-in-class cardiolipin-protective compound as a therapeutic agent to restore mitochondrial bioenergetics. Br J Pharmacol 171:2029–2050
113. Myers L, Farmer JM, Wilson RB, Friedman L, Tsou A, Perlman SL, Subramony SH, Gomez CM, Ashizawa T, Wilmot GR, Mathews KD, Balcer LJ, Lynch DR (2008) Antioxidant use in Friedreich ataxia. J Neurol Sci 267:174–176

Chapter 8
The Role of Iron in Amyotrophic Lateral Sclerosis

Xian-Le Bu, Yang Xiang and Yansu Guo

Abstract Amyotrophic lateral sclerosis (ALS) is a progressive neurodegenerative disorder characterized by the degeneration of motor neurons in the motor cortex, brainstem, and spinal cord. The etiology and pathogenesis of this devastating disease remain largely unknown. Increasing evidence suggests that iron accumulation is involved in the onset and progression of ALS. In this review, we discuss the regulation of iron homoeostasis in the brain, the misregulation of iron homeostasis in ALS, and its possible roles in the mechanism of the disease. Finally, we summarize the recent progress and problems with respect to iron chelator therapies on ALS, aiming to propose a new therapeutic strategy to ameliorate the progression of the disease.

Keywords Amyotrophic lateral sclerosis · Pathogenesis · Iron · Iron chelator

8.1 Introduction

Amyotrophic lateral sclerosis (ALS) is a progressive and devastating neurodegenerative disorder characterized with the degeneration of upper and lower motor neurons in the spinal anterior horn and motor cortex and loss of axons in the lateral columns of the spinal cord [1]. ALS is a clinically heterogeneous disease as patients present a large variability in age, onset site, initial symptoms, and speed of disease progression. Almost 90% of ALS patients are sporadic form and the rest are familial form caused by genetic mutations. However, the clinical manifestations and the pathological hallmarks are similar, and no effective treatment is available now. The pathogenesis of this fatal neurodegenerative disease remains largely unknown. Numerous evidences

X.-L. Bu
Department of Neurology and Centre for Clinical Neuroscience, Daping Hospital, Third Military Medical University, Chongqing, China

Y. Xiang
Department of Neurology, Chengdu Military General Hospital, Chengdu, China

Y. Guo (✉)
Beijing Geriatric Healthcare Center, Xuanwu Hospital, Capital Medical University, Beijing, China
e-mail: gys188@163.com

Y.-Z. Chang (ed.), *Brain Iron Metabolism and CNS Diseases*, Advances in Experimental Medicine and Biology 1173,
https://doi.org/10.1007/978-981-13-9589-5_8

suggest that redox metals, mitochondrial dysfunction, oxidative stress, and neuroinflammation are involved in the loss of motor neurons [2–4].

Iron has been proved to play a crucial role in oxygen transportation, cell proliferation and differentiation, DNA synthesis, myelin production, synthesis of neurotransmitters, and the generation of oxidative stress [5]. Many studies indicate that iron overload may be involved in ALS pathogenesis. In this review, we summarize the current knowledge of iron metabolism disturbances and its pathogenic role in ALS, and discuss the recent progress and problems with respect to iron chelator therapy on ALS, aiming to propose a new therapeutic strategy to ameliorate the progression of the disease.

8.2 Iron Homoeostasis in the Brain

Transferrin receptor 1 (TFR1) is responsible for the endocytosis of the iron uptake, and it is highly expressed on the luminal side of endothelial cells [6]. Blood iron enters the endothelial cells of the blood–brain barrier via TFR1-mediated endocytosis [7–9]. Iron could also be released into the extracellular environment and then be taken up by neurons, microglia, and astrocytes [5, 10].

Neurons are thought to uptake iron via the transferrin receptors and divalent metal transporter 1 (DMT 1). Meanwhile, neurons express ferroportin, which is involved in the efflux of iron [11]. It has been shown that inflammatory stimuli could increase DMT1 expression while decrease the expression of ferroportin [12], causing the iron accumulation in neurons. Astrocyte, the important component of the blood-brain barrier, plays a key role in regulating brain iron absorption [13]. Although astrocytes do not express TFR, it is able to take up iron via the DMT1 that expressed in the perivascular end-foot processes of astrocytes [14]. The glycosylphosphatidylinositol-anchored form of ceruloplasmin expressed by astrocytes plays a role in oxidizing ferrous iron, and could help iron efflux from cells [15]. It shows that lack of ceruloplasmin can lead to iron accumulation in the brain and neurodegeneration [16]. Iron is required for axon myelination in oligodendrocytes. It has been proved that oligodendrocytes can take up interstitial ferritin via the ferritin receptor. On the other hand, oligodendrocyte is able to uptake iron from blood vessels [5]. Inflammatory stimuli increases iron accumulation in microglia by increasing the expression of DMT1 protein [17, 18]. It showed that iron uptake was increased and iron exportation was down-regulated when the microglia is activated [19]. Moreover, inflammatory stimuli could increase hepcidin concentrations in astrocytes and microglia, and decrease ferroportin level in astrocytes, microglia, and neurons [12].

Neuroinflammation is a chronic activation of the immune response in the central nervous system and related to neuronal dysfunction in neurodegenerative diseases. Microglia could be switched from their normal quiescent state to an activated state by the systemic inflammation with aging [20]. Age-dependent enhanced neuroinflammatory processes in the brain refer to increased microglia and astrocyte numbers and proinflammatory cytokines secreted by activated glial cells with aging [20]. Iron levels in the substantia nigra, putamen, globus pallidus, caudate nucleus, and cortices increase with age [5]. Additionally, the blood–brain barrier permeability increases in the process of aging. Therefore, brain iron load may increase during aging.

8.3 Misregulation of Iron Homeostasis in Amyotrophic Lateral Sclerosis

Numerous studies have examined the iron levels in ALS patients. A prior study reported in 1985 showed that the iron concentrations of erythrocytes were the same between ALS and controls [21]. Goodall et al. declared that the serum ferritin levels were significantly increased in ALS patients, while no significant differences in the concentrations of iron, transferrin, iron saturation, or total iron binding capacity were found [22]. It also reported that ALS patients had significantly higher plasma levels of ferritin but lower concentrations of transferrin [23, 24], and the patients with a higher ferritin level had a shorter survival time [25]. A recent study firstly showed that the serum iron, ferritin, and transferrin saturation coefficient were significantly elevated in ALS patients, and the iron status was associated with body weight loss [26]. And serum ferritin may be a candidate biomarker for ALS aggravation [27]. Free iron level was also increased in the cerebrospinal fluid of ALS patients compared to controls [28, 29]. And twofold increased level of inappropriate iron ligands in cerebrospinal fluid (CSF) are found in patients with ALS, which may increase iron redox activity and reactive oxygen species (ROS) production [30].

Autopsy study demonstrated that iron load in gray matter from the frontal cortex of ALS was increased significantly than that of controls [31]. The concentration of iron was also increased in the spinal cord of ALS patients [32–34]. Disruption in the expression of brain iron transporters such as lactotransferrin receptor, melanotransferrin, and ceruloplasmin is related to iron accumulation in ALS [35]. Iron accumulation could also be observed in the motor cortex using magnetic resonance imaging (MRI), which appears to be a very promising biomarker for ALS [36, 37]. MRI also found showed increased iron deposits in the globus pallidus of frontotemporal dementia with ALS [38]. 7T MRI showed that ALS patients had motor cortex hypointensity, and the postmortem studies showed that the increased iron accumulation in microglial cells may be responsible for the location of the signal changes on MRI [39]. Recent researches confirmed that iron level was significantly higher in the motor cortex of ALS patients by susceptibility-weighted imaging (SWI), which may be useful biomarkers for the diagnosis of clinically suspected ALS [40, 41].

8.4 The Link of Iron with Amyotrophic Lateral Sclerosis

Iron is a redox ion and can aggravate oxidative stress by generating ROS, which may lead to neuronal cell damage, aging, and apoptosis. It demonstrated that the increased iron load can form redox active complexes which participates in the Fenton reaction and produces the toxic hydroxyl radical [42–44]. Meanwhile, ROS can induce the release of iron from the iron storage proteins such as mitochondrial iron—sulfur cluster proteins [45]. Increased oxidative stress occurred in ALS could also be promoted by the genetic abnormalities (e.g., superoxide dismutase mutations) [46]. Iron accumulation also increases the enzymatic activity of TNF-α converting enzyme, leading to secretion of TNF-α and promoting neuroinflammation [47]. A number of studies have demonstrated that mutations of genes encoding the proteins involved in iron regulation are related to neuronal cell death also in ALS [48, 49]. The mutation of HFE gene, which encoding protein interacts with the transferrin receptor and regulates transferrin-mediated iron uptake, altered brain iron profiles, increased oxidative stress, and accelerated disease process in mice [50, 51]. The mutation of SLC11A2 gene that encoding DMT1 is associated with shorter disease duration in ALS cases [52]. On the other hand, activation of genes involved in iron homeostasis can induce ferritin overexpression and thus limit toxic hydroxyl radical and neuronal loss [53]. Thus, iron misregulation promotes oxidative stress, which disrupts cellular iron balance and causes progressive loss of motor neurons and consequent oxidative stress, thus giving rise to a vicious circle.

8.5 Iron Chelator Therapy on Amyotrophic Lateral Sclerosis

In view of the key role of iron in the pathogenesis of ALS, researchers explore the efficacy and therapeutic potential of brain-permeable iron-chelating drugs for ALS in animal models. An earlier study found that intraperitoneal injection of iron porphyrin significantly decreased neuronal death and improved motor function in G93A mutant SOD1 transgenic mouse model of ALS [54]. VAR10303, the brain-permeable iron-chelating radical scavenging drug, exerted several beneficial effects in $SOD1^{G93A}$ transgenic ALS mice, including alleviating the iron accumulation and motoneuron loss in the spinal cord, improving motor performance and prolonging survival time [55]. Iron-chelating drugs M30 and HLA20 showed neuroprotective effects against oxidative stress and can extend the survival time of ALS transgenic mice [56]. Iron chelators VK-28 also significantly alleviated the increased levels of iron and oxygen free radicals and suppressed glial activation in the spinal cords of the ALS mice [57]. What's more, it was able to decrease TDP-43 protein aggregation and the proapoptotic molecule Bax and to increase the expression of antiapoptotic protein Bcl-2 [57]. A recent pilot clinical trial showed that deferiprone could remarkably decrease the iron levels in the cervical spinal cord, medulla oblongata, and motor cortex, the levels of oxidative stress markers and neurofilament light chains in the CSF, and attenuate the disability progression and weight loss in patients with ALS [58]. Therefore, brain-permeable iron-chelating drugs hold a promise as therapeutic agents for ALS by targeting multiple key pathways of the disease pathogenesis. And the effectiveness of iron chelators in ALS therapy supports the role of iron in the ALS pathogenesis.

8.6 Conclusion

In summary, abundant evidence shows that iron accumulation in motor cortex and spinal cord is a hallmark of ALS, and increased iron load, in particular, in regions of central nervous system causes neuronal cell damage. However, it is uncertain whether iron accumulation is a primary cause of ALS or consequential pathological change. And why iron accumulation in ALS is selective for some brain areas is also unclear. To further clarify the pathophysiological role of iron in ALS pathogenesis, future studies should focus on the chemical state of iron in healthy control and ALS patients, and its specifically pathogenic role in ALS different from other neurodegenerative disorders.

More randomized, double-blind, placebo-controlled, multicenter trials are needed to evaluate the effect of iron chelators on patients with ALS.

References

1. Vandoorne T, De Bock K, Van Den Bosch L (2018) Energy metabolism in ALS: an underappreciated opportunity? Acta Neuropathol 135:489–509
2. Carri MT, Ferri A, Cozzolino M, Calabrese L, Rotilio G (2003) Neurodegeneration in amyotrophic lateral sclerosis: the role of oxidative stress and altered homeostasis of metals. Brain Res Bull 61:365–374
3. Hadzhieva M, Kirches E, Wilisch-Neumann A, Pachow D, Wallesch M, Schoenfeld P, Paege I, Vielhaber S, Petri S, Keilhoff G, Mawrin C (2013) Dysregulation of iron protein expression in the G93A model of amyotrophic lateral sclerosis. Neuroscience 230:94–101
4. Tokuda E, Furukawa Y (2016) Copper homeostasis as a therapeutic target in amyotrophic lateral sclerosis with SOD1 mutations. Int J Mol Sci 17(5):636
5. Ward RJ, Zucca FA, Duyn JH, Crichton RR, Zecca L (2014) The role of iron in brain ageing and neurodegenerative disorders. Lancet Neurol 13:1045–1060
6. Jefferies WA, Brandon MR, Hunt SV, Williams AF, Gatter KC, Mason DY (1984) Transferrin receptor on endothelium of brain capillaries. Nature 312:162–163
7. Moos T, Rosengren Nielsen T, Skjorringe T, Morgan EH (2007) Iron trafficking inside the brain. J Neurochem 103:1730–1740
8. Ke Y, Qian ZM (2007) Brain iron metabolism: neurobiology and neurochemistry. Prog Neurobiol 83:149–173
9. Leitner DF, Connor JR (1820) Functional roles of transferrin in the brain. Biochim Biophys Acta 2012:393–402
10. Biasiotto G, Di Lorenzo D, Archetti S, Zanella I (2016) Iron and neurodegeneration: is ferritinophagy the link? Mol Neurobiol 53:5542–5574

11. Boserup MW, Lichota J, Haile D, Moos T (2011) Heterogenous distribution of ferroportin-containing neurons in mouse brain. Biometals 24:357–375
12. Urrutia P, Aguirre P, Esparza A, Tapia V, Mena NP, Arredondo M, Gonzalez-Billault C, Nunez MT (2013) Inflammation alters the expression of DMT1, FPN1 and hepcidin, and it causes iron accumulation in central nervous system cells. J Neurochem 126:541–549
13. Dringen R, Bishop GM, Koeppe M, Dang TN, Robinson SR (2007) The pivotal role of astrocytes in the metabolism of iron in the brain. Neurochem Res 32:1884–1890
14. Abbott NJ, Ronnback L, Hansson E (2006) Astrocyte-endothelial interactions at the blood-brain barrier. Nat Rev Neurosci 7:41–53
15. Patel BN, David S (1997) A novel glycosylphosphatidylinositol-anchored form of ceruloplasmin is expressed by mammalian astrocytes. J Biol Chem 272:20185–20190
16. Jeong SY, David S (2003) Glycosylphosphatidylinositol-anchored ceruloplasmin is required for iron efflux from cells in the central nervous system. J Biol Chem 278:27144–27148
17. Nemeth E, Rivera S, Gabayan V, Keller C, Taudorf S, Pedersen BK, Ganz T (2004) IL-6 mediates hypoferremia of inflammation by inducing the synthesis of the iron regulatory hormone hepcidin. J Clin Invest 113:1271–1276
18. Lee P, Peng H, Gelbart T, Beutler E (2004) The IL-6- and lipopolysaccharide-induced transcription of hepcidin in HFE-, transferrin receptor 2-, and beta 2-microglobulin-deficient hepatocytes. Proc Natl Acad Sci U S A 101:9263–9265
19. Xu H, Wang Y, Song N, Wang J, Jiang H, Xie J (2017) New progress on the role of glia in iron metabolism and iron-induced degeneration of dopamine neurons in Parkinson's disease. Front Mol Neurosci 10:455
20. Liu YH, Zeng F, Wang YR, Zhou HD, Giunta B, Tan J, Wang YJ (2013) Immunity and Alzheimer's disease: immunological perspectives on the development of novel therapies. Drug Discovery Today 18:1212–1220
21. Nagata H, Miyata S, Nakamura S, Kameyama M, Katsui Y (1985) Heavy metal concentrations in blood cells in patients with amyotrophic lateral sclerosis. J Neurol Sci 67:173–178
22. Goodall EF, Haque MS, Morrison KE (2008) Increased serum ferritin levels in amyotrophic lateral sclerosis (ALS) patients. J Neurol 255:1652–1656
23. Mitchell RM, Simmons Z, Beard JL, Stephens HE, Connor JR (2010) Plasma biomarkers associated with ALS and their relationship to iron homeostasis. Muscle Nerve 42:95–103
24. Qureshi M, Brown RH Jr, Rogers JT, Cudkowicz ME (2008) Serum ferritin and metal levels as risk factors for amyotrophic lateral sclerosis. Open Neurol J 2:51–54
25. Nadjar Y, Gordon P, Corcia P, Bensimon G, Pieroni L, Meininger V, Salachas F (2012) Elevated serum ferritin is associated with reduced survival in amyotrophic lateral sclerosis. PLoS ONE 7:e45034
26. Veyrat-Durebex C, Corcia P, Mucha A, Benzimra S, Mallet C, Gendrot C, Moreau C, Devos D, Piver E, Pages JC, Maillot F, Andres CR, Vourc'h P, Blasco H (2014) Iron metabolism disturbance in a French cohort of ALS patients. Biomed Res Int 2014:485723
27. Yu J, Wang N, Qi F, Wang X, Zhu Q, Lu Y, Zhang H, Che F, Li W (2018) Serum ferritin is a candidate biomarker of disease aggravation in amyotrophic lateral sclerosis. Biomed Rep 9:333–338
28. Kokic AN, Stevic Z, Stojanovic S, Blagojevic DP, Jones DR, Pavlovic S, Niketic V, Apostolski S, Spasic MB (2005) Biotransformation of nitric oxide in the cerebrospinal fluid of amyotrophic lateral sclerosis patients. Redox Rep 10:265–270
29. Hozumi I, Hasegawa T, Honda A, Ozawa K, Hayashi Y, Hashimoto K, Yamada M, Koumura A, Sakurai T, Kimura A, Tanaka Y, Satoh M, Inuzuka T (2011) Patterns of levels of biological metals in CSF differ among neurodegenerative diseases. J Neurol Sci 303:95–99
30. Ignjatovic A, Stevic Z, Lavrnic D, Nikolic-Kokic A, Blagojevic D, Spasic M, Spasojevic I (2012) Inappropriately chelated iron in the cerebrospinal fluid of amyotrophic lateral sclerosis patients. Amyotroph Lateral Scler 13:357–362
31. Yasui M, Ota K, Garruto RM (1993) Concentrations of zinc and iron in the brains of Guamanian patients with amyotrophic lateral sclerosis and parkinsonism-dementia. Neurotoxicology 14:445–450

32. Ince PG, Shaw PJ, Candy JM, Mantle D, Tandon L, Ehmann WD, Markesbery WR (1994) Iron, selenium and glutathione peroxidase activity are elevated in sporadic motor neuron disease. Neurosci Lett 182:87–90
33. Kasarskis EJ, Tandon L, Lovell MA, Ehmann WD (1995) Aluminum, calcium, and iron in the spinal cord of patients with sporadic amyotrophic lateral sclerosis using laser microprobe mass spectroscopy: a preliminary study. J Neurol Sci 130:203–208
34. Markesbery WR, Ehmann WD, Candy JM, Ince PG, Shaw PJ, Tandon L, Deibel MA (1995) Neutron activation analysis of trace elements in motor neuron disease spinal cord. Neurodegeneration 4:383–390
35. Qian ZM, Wang Q (1998) Expression of iron transport proteins and excessive iron accumulation in the brain in neurodegenerative disorders. Brain Res Brain Res Rev 27:257–267
36. Ignjatovic A, Stevic Z, Lavrnic S, Dakovic M, Bacic G (2013) Brain iron MRI: a biomarker for amyotrophic lateral sclerosis. J Magn Reson Imaging 38:1472–1479
37. Oba H, Araki T, Ohtomo K, Monzawa S, Uchiyama G, Koizumi K, Nogata Y, Kachi K, Shiozawa Z, Kobayashi M (1993) Amyotrophic lateral sclerosis: T2 shortening in motor cortex at MR imaging. Radiology 189:843–846
38. Santillo AF, Skoglund L, Lindau M, Eeg-Olofsson KE, Tovi M, Engler H, Brundin RM, Ingvast S, Lannfelt L, Glaser A, Kilander L (2009) Frontotemporal dementia-amyotrophic lateral sclerosis complex is simulated by neurodegeneration with brain iron accumulation. Alzheimer Dis Assoc Disord 23:298–300
39. Kwan JY, Jeong SY, Van Gelderen P, Deng HX, Quezado MM, Danielian LE, Butman JA, Chen L, Bayat E, Russell J, Siddique T, Duyn JH, Rouault TA, Floeter MK (2012) Iron accumulation in deep cortical layers accounts for MRI signal abnormalities in ALS: correlating 7 tesla MRI and pathology. PLoS ONE 7:e35241
40. Yu J, Qi F, Wang N, Gao P, Dai S, Lu Y, Su Q, Du Y, Che F (2014) Increased iron level in motor cortex of amyotrophic lateral sclerosis patients: an in vivo MR study. Amyotroph Lateral Scler Frontotemporal Degener 15:357–361
41. Adachi Y, Sato N, Saito Y, Kimura Y, Nakata Y, Ito K, Kamiya K, Matsuda H, Tsukamoto T, Ogawa M (2015) Usefulness of SWI for the detection of iron in the motor cortex in amyotrophic lateral sclerosis. J Neuroimaging 25:443–451
42. Popovic-Bijelic A, Mojovic M, Stamenkovic S, Jovanovic M, Selakovic V, Andjus P, Bacic G (2016) Iron-sulfur cluster damage by the superoxide radical in neural tissues of the SOD1(G93A) ALS rat model. Free Radic Biol Med 96:313–322
43. Sheykhansari S, Kozielski K, Bill J, Sitti M, Gemmati D, Zamboni P, Singh AV (2018) Redox metals homeostasis in multiple sclerosis and amyotrophic lateral sclerosis: a review. Cell Death Dis 9:348
44. Bozzo F, Mirra A, Carri MT (2017) Oxidative stress and mitochondrial damage in the pathogenesis of ALS: new perspectives. Neurosci Lett 636:3–8
45. Horowitz MP, Greenamyre JT (2010) Mitochondrial iron metabolism and its role in neurodegeneration. J Alzheimers Dis 20(Suppl 2):S551–S568
46. Mattson MP (2004) Metal-catalyzed disruption of membrane protein and lipid signaling in the pathogenesis of neurodegenerative disorders. Ann N Y Acad Sci 1012:37–50
47. Lee JK, Shin JH, Gwag BJ, Choi EJ (2015) Iron accumulation promotes TACE-mediated TNF-alpha secretion and neurodegeneration in a mouse model of ALS. Neurobiol Dis 80:63–69
48. Sirabella R, Valsecchi V, Anzilotti S, Cuomo O, Vinciguerra A, Cepparulo P, Brancaccio P, Guida N, Blondeau N, Canzoniero LMT, Franco C, Amoroso S, Annunziato L, Pignataro G (2018) Ionic homeostasis maintenance in ALS: focus on new therapeutic targets. Front Neurosci 12:510
49. Gajowiak A, Stys A, Starzynski RR, Staron R, Lipinski P (2016) Misregulation of iron homeostasis in amyotrophic lateral sclerosis. Postepy Hig Med Dosw (Online) 70:709–721
50. Nandar W, Neely EB, Unger E, Connor JR (1832) A mutation in the HFE gene is associated with altered brain iron profiles and increased oxidative stress in mice. Biochim Biophys Acta 2013:729–741

51. Nandar W, Neely EB, Simmons Z, Connor JR (1842) H63D HFE genotype accelerates disease progression in animal models of amyotrophic lateral sclerosis. Biochim Biophys Acta 2014:2413–2426
52. Blasco H, Vourc'h P, Nadjar Y, Ribourtout B, Gordon PH, Guettard YO, Camu W, Praline J, Meininger V, Andres CR, Corcia P (2011) Association between divalent metal transport 1 encoding gene (SLC11A2) and disease duration in amyotrophic lateral sclerosis. J Neurol Sci 303:124–127
53. Olsen MK, Roberds SL, Ellerbrock BR, Fleck TJ, McKinley DK, Gurney ME (2001) Disease mechanisms revealed by transcription profiling in SOD1-G93A transgenic mouse spinal cord. Ann Neurol 50:730–740
54. Wu AS, Kiaei M, Aguirre N, Crow JP, Calingasan NY, Browne SE, Beal MF (2003) Iron porphyrin treatment extends survival in a transgenic animal model of amyotrophic lateral sclerosis. J Neurochem 85:142–150
55. Golko-Perez S, Amit T, Bar-Am O, Youdim MB, Weinreb O (2017) A novel iron chelator-radical scavenger ameliorates motor dysfunction and improves life span and mitochondrial biogenesis in SOD1(G93A) ALS Mice. Neurotox Res 31:230–244
56. Kupershmidt L, Weinreb O, Amit T, Mandel S, Carri MT, Youdim MB (2009) Neuroprotective and neuritogenic activities of novel multimodal iron-chelating drugs in motor-neuron-like NSC-34 cells and transgenic mouse model of amyotrophic lateral sclerosis. FASEB J 23:3766–3779
57. Wang Q, Zhang X, Chen S, Zhang X, Zhang S, Youdium M, Le W (2011) Prevention of motor neuron degeneration by novel iron chelators in SOD1(G93A) transgenic mice of amyotrophic lateral sclerosis. Neurodegener Dis 8:310–321
58. Moreau C, Danel V, Devedjian JC, Grolez G, Timmerman K, Laloux C, Petrault M, Gouel F, Jonneaux A, Dutheil M, Lachaud C, Lopes R, Kuchcinski G, Auger F, Kyheng M, Duhamel A, Perez T, Pradat PF, Blasco H, Veyrat-Durebex C, Corcia P, Oeckl P, Otto M, Dupuis L, Garcon G, Defebvre L, Cabantchik ZI, Duce J, Bordet R, Devos D (2018) Could conservative iron chelation lead to neuroprotection in amyotrophic lateral sclerosis? Antioxid Redox Signal 29:742–748

Chapter 9
Iron Pathophysiology in Neurodegeneration with Brain Iron Accumulation

Sonia Levi, Anna Cozzi and Paolo Santambrogio

Abstract Neurodegeneration with brain iron accumulation (NBIA) is a group of seriously devastating and life-threatening rare monogenic diseases characterized by focal iron accumulation in the brain. The main symptoms of NBIA comprise progressive movement disorder, often including painful dystonia, parkinsonism, mental disability, and early death. Currently, a single established therapy is not available to reverse the progression of these debilitating disorders. The complexity of NBIA emerged from the identification of various causative genes, and up to 15 genes have been identified to date. Although the NBIA genes are involved in different cellular biochemical pathways, they show the common characteristic of generating severe iron accumulation in the basal ganglia of the patients' brains. Thus, the molecular events that lead to brain iron overload and their important roles in the pathophysiology of the diseases are not easy to identify and are poorly understood. This review summarizes the current knowledge on NBIA disorders, with a particular focus on the data describing the role of iron in the pathogenic mechanisms.

Keywords Iron · Neurodegenerative disorders · NBIA

9.1 Introduction

Neurodegeneration with brain iron accumulation (NBIA) comprises a wide spectrum of clinically and genetically heterogeneous diseases (Table 9.1). The heterogeneity is due to the variability in the onset of clinical symptoms that are recapitulated in dystonia, pyramidal and extrapyramidal signs, cognitive decline, behavioral abnormalities, and retinal and axonal neuropathy [47]. Nevertheless, these diseases are classified as a unique group of disorders because they are characterized by iron overload in

S. Levi (✉)
Division of Neuroscience, San Raffaele Scientific Institute, Vita-Salute San Raffaele University, 20132 Milan, Italy
e-mail: levi.sonia@hsr.it

A. Cozzi · P. Santambrogio
Division of Neuroscience, San Raffaele Scientific Institute, 20132 Milan, Italy

Y.-Z. Chang (ed.), *Brain Iron Metabolism and CNS Diseases*,
Advances in Experimental Medicine and Biology 1173,
https://doi.org/10.1007/978-981-13-9589-5_9

Table 9.1 Neurodegeneration with brain iron accumulation disorders

Gene	Function	Inheritance	Human disorder	Brain iron	% in NBIA
FLT1; L-Ferritin (ferritin light chain)	Iron storage	AD	Neuroferritinopathy, hereditary ferritinopathy, NBIA type III	Dense ferritin-Fe spheroid inclusions throughout neurons, oligodendrocytes, and microglia in the forebrain and cerebellum	Rare
Cp; Ceruloplasmin	Cu-dependent-ferroxidase	AR	Aceruloplasminemia	Iron accumulation in the basal ganglia	Rare
PANK2; Pantothenate kinase-2	Coenzyme A synthesis and fatty acid metabolism	AR	PKAN, NBIA type I, Hallervorden–Spatz syndrome	Iron accumulation in the basal ganglia (eye of the tiger)	35–50
COASY	CoA synthase	AR	CoPAN, COASY protein-associated neurodegeneration	Iron deposition in the globus pallidus and substantia nigra	Rare
PLA2G6; iPLA2;	Arachidonic acid release	AR	PLAN, NBIA type II, INAD1, Karak syndrome	Iron deposition in the globus pallidus and substantia nigra (in <50% of patients)	20
C19orf12	Unknown, lipid metabolism?	AR	MPAN, mitochondrial membrane-associated neurodegeneration	Iron deposition in the globus pallidus and substantia nigra	6–10
FA2H	Fatty acid hydroxylase	AR	FAHN, fatty acid hydroxylase-associated neurodegeneration	Iron deposition in the globus pallidus and substantia nigra in some cases	Rare
WDR45	Autophagy	XLD	BPAN, β-propeller protein-associated neurodegeneration	Predominant iron deposition in the globus pallidus compared with the substantia nigra	1–2

(continued)

Table 9.1 (continued)

Gene	Function	Inheritance	Human disorder	Brain iron	% in NBIA
SCP2	Peroxisomal enzyme	AR	Leukoencephalopathy with dystonia and motor neuropathy	Iron deposition in the thalamus	Rare
CRAT	Carnitine acetyl transferase, β-oxidation	AR	*CRAT*-associated neurodegeneration	Iron accumulation in the globus pallidus and substantia nigra	Rare
AP4M1	Vesicle formation	AR	*AP4M1*-associated neurodegeneration	Iron in globus pallidus in a single family	Rare
REPS1	Endocytosis and vesicle transport	AR	*RESP1*-associated neurodegeneration	Iron accumulation in the globus pallidus and peduncles.	Rare
ATP13A2	Lysosomal cation pump, autophagosome formation	AR	Kufor–Rakeb syndrome	Few patients show iron deposition in the striatum	Rare
DCAF17 (C2orf37)	Nucleolar protein	AR	Woodhouse–Sakati syndrome	Iron deposition in the globus pallidus and substantia nigra is only detected in some patients	Rare
GTPBP2	Unknown, mRNA/ribosome stability?	AR	*GTPBP2*-associated neurodegeneration	Iron overload in the globus pallidus and substantia nigra	Rare

specific brain areas, mainly in the basal ganglia, even if this feature is not present in all patients [71]. The brain iron burden has been visualized in vivo using MRI [30], and in several cases, it was also confirmed by a histopathological examinations in postmortem brain samples [64]. Indeed, MRI is the main method used to analyze areas of iron accumulation in the brain, and this feature, combined with the variations in the structures of certain areas, is the basis of distinguishing between types of NBIA [79]. Thus, the radiological evidence of iron accumulation together with the clinical symptoms is the preliminary diagnostic tool that suggests to clinicians to undertake a screening to identify the specific causative gene [99]. Currently, 15 genes have been associated with different subtypes of NBIA, the main characteristics of which are described in Table 9.1. Among these diseases, only two, neuroferritinopathy and aceruloplasminemia, are linked to mutations in genes that are directly involved in iron metabolism, *FTL1* and *CP*, respectively. All the other forms (PKAN, CoPAN, PLAN, MPAN, FAHN, BPAN, leukoencephalopathy with dystonia and motor neuropathy, Kufor–Rakeb syndrome, Woodhouse–Sakati syndrome, and GTPBP2-, CRAT-, REPS1-, and AP4M1-associated neurodegeneration) are characterized by altered iron metabolism and its accumulation in the brain via an as-yet-unidentified mechanism, possibly implicating other pathways. These disorders are indeed caused by mutations in genes linked to lipid metabolism (PLAN, FAHN and leukoencephalopathy with dystonia and motor neuropathy), mitochondrial activities (PKAN, CoPAN, MPAN and CRAT-associated neurodegeneration), autophagy (BPAN, Kufor–Rakeb syndrome and REPS1- and AP4M1-associated neurodegeneration) and some still unknown functions (Woodhouse–Sakati syndrome, GTPBP2-associated neurodegeneration), all of which are apparently unrelated to iron metabolism (Table 9.1).

Interestingly, numerous genes associated with inherited NBIA encode proteins localized in the mitochondria (PANK2, COPAN, PLA2G6, C19orf12, ATP13A2, and CRAT). Thus, these pathologies represent an important opportunity to study and understand the molecular mechanisms involved in brain iron management and the connections between mitochondria function and neurodegeneration. In this review, we focus on the reported data regarding iron alterations, with the aim of highlighting the knowledge of the involvement of this metal in the pathogenesis of NBIA disorders.

9.2 Fe-Related NBIA

Aceruloplasminemia, which is associated with recessive mutations in *CP*, and Neuroferritinopathy, which is linked to mutations in *FTL1*, are the two NBIA subtypes caused by proteins that are important in maintaining iron homeostasis and therefore have been extensively investigated in this context.

9.2.1 Aceruloplasminemia

Aceruloplasminemia is the first disease in which the alteration of an iron protein has been associated with neurodegeneration (OMIM #604290). *CP* is located on chromosome 3q 21–24 and encodes ceruloplasmin (Cp), a protein with ferroxidase function. Aceruloplasminemia is an autosomal recessive inherited disease that was originally described in a Japanese female in 1987 [82]. Indeed, it particularly affects the Japanese population, with a prevalence of approximately one per 2,000,000. A genetic analysis of patients affected by aceruloplasminemia revealed approximately 50 distinct causative mutations identified in more than 60 families worldwide [60, 81].

The symptoms include neurological signs such as blepharospasm, grimacing, facial and neck dystonia, tremors, chorea, and ataxia, with the first appearance in adulthood (fourth or fifth decade of life). Cognitive dysfunction is also observed in a large portion of affected subjects. Generally, the neurological symptoms are preceded by diabetes mellitus, retinal degeneration, and microcytic anemia [59]. The other hematological parameters displaying alterations include undetectable serum ceruloplasmin levels, high serum ferritin levels, and low levels of sideremia [85]. The striatum, thalamus, and dentate nucleus are the regions of the brain affected by iron deposition, as revealed by abnormally low intensities on T1- and T2-weighted MRIs [80]. Neuropathology indicates a severe destruction of the basal ganglia and dentate nucleus, with considerable iron deposition in neuronal and glial cells [38, 56, 114]. Iron was localized to terminal astrocytic processes, and, accordingly, enlarged or deformed astrocytes and spheroid-like globular structures are characteristic neuropathological findings in patients with aceruloplasminemia [38]. Iron overload was also detected in the cerebral cortex [56], retina and cerebellum, as well as in the pancreas and myocardium [56].

Cp is a glycoprotein present in the β2-globulin fraction of serum. It is a multicopper ferroxidase, containing 95% of the copper in the plasma. In the CNS, Cp is expressed as a glycosylphosphatidylinositol (GPI)-linked form in astrocytes [89]. Its function is to facilitate the export of iron mediated by ferroportin, oxidizing Fe^{2+} to Fe^{3+} and facilitating Fe^{3+} binding to the transferrin present in the extracellular environment. This action is essential in astrocytes, for which ceruloplasmin is the only existing ferroxidase [55]. *CP* mutations cause structural modifications in the protein that induces the retention of Cp in the ER, miss-incorporation of copper into Cp or impaired ferroxidase activity [27, 48, 62, 63]. In all these cases, iron export from the cell is impaired, leading to cellular iron overload.

The molecular pathogenesis of aceruloplasminemia was examined by analyzing mammalian cell cultures expressing *CP* mutants [27, 48, 62, 63] and by characterizing several different murine models [45, 89, 106, 112]. The overall data collected from these models indicated that when the function of Cp is altered, the iron entering in the CNS as ferrous iron is not oxidized. Cells exposed to excess ferrous iron internalize it in large quantities through non-regulated internalization pathways of non-transferrin-bound iron [14]. The excess iron import combined with the inability of ferroportin to export iron caused by the absence of Cp leads to the accumulation of the metal in astrocytes, as observed in patients. Iron may remain in astrocytes and does not reach the neurons, which then die due to the iron deficiency and from exposure to toxins released by dying astrocytes. Nevertheless, some mutated forms of *CP* also accumulate in aggregates and lead to the death of astrocytes through an iron-independent pathway [61].

Other cells in the CNS, including oligodendrocytes, do not depend on the ferro-oxidative function of Cp, but require the function of hephaestin; this finding explains the specificity of astrocytes and neuronal death. Only the contemporary ablation of the CP and hephaestin in mice effectively produces symptoms consistent with patients with aceruloplasminemia [44, 103]. These animals exhibit iron accumulation in oligodendrocytes present in both the gray and white matter [103], macular degeneration, iron overload, and increased oxidative stress in the retina [42]. In this animal model, the administration of the oral iron chelator deferiprone protected against the increase in oxidative stress, retinal degeneration, and lipofuscin accumulation and promoted a reduction in ataxia and an increase in lifespan [43]. Despite the good results obtained from mice models, the administer iron chelation therapies to patients partially controls systemic iron deposition but was ineffective against neurodegeneration [56, 96]. However, iron-chelating therapy appeared to be effective in reducing the hepatic and pancreatic iron overload [35]. Another therapeutic approach is ceruloplasmin replacement therapy. The parental administration of human ceruloplasmin to *Cp-/-* mice restored the protein levels and ferroxidase activity in the brain in a recent study [115]. Interestingly, in ceruloplasmin-treated mice, reduced brain iron deposition was associated with the amelioration of impairments in motor coordination, indicating that enzyme replacement therapy may be a promising strategy [115]. This finding further supports the overall data collected, indicating that oxidative stress driven by heavy metal accumulation represents the primary cellular cytotoxic process that determines the neuronal damage in affected brain regions.

9.2.2 Neuroferritinopathy

Neuroferritinopathy (NF, OMIM#606159) is an autosomal dominant inherited disease linked to mutations in *FTL1*. The gene product is L-ferritin subunit that complexes with the H-ferritin subunit to form the major iron storage protein. NF was first described in 2001 in a large English family carrying an adenine insertion at position 460–461 (c.460dupA) of the *FTL1* gene [22]. After this first identification, 10 different mutations (Fig. 9.1) were described in families from different parts of the world [22, 26, 66, 76, 83, 84, 86, 105, 108]. The peculiarity of these DNA alterations is the restricted area of the gene in which they arise. In fact, 9 of these causative mutations are duplications of one or more nucleotides in exon 4 of the gene. All these variants cause a frameshift of different lengths in the sequence that modifies the C-terminal sequence of the L-peptide (Fig. 9.1) [72]. Moreover, the missense Ala96Thr mutation in intron 3 that modifies the middle of the *FTL1* sequence is associated with NF [74].

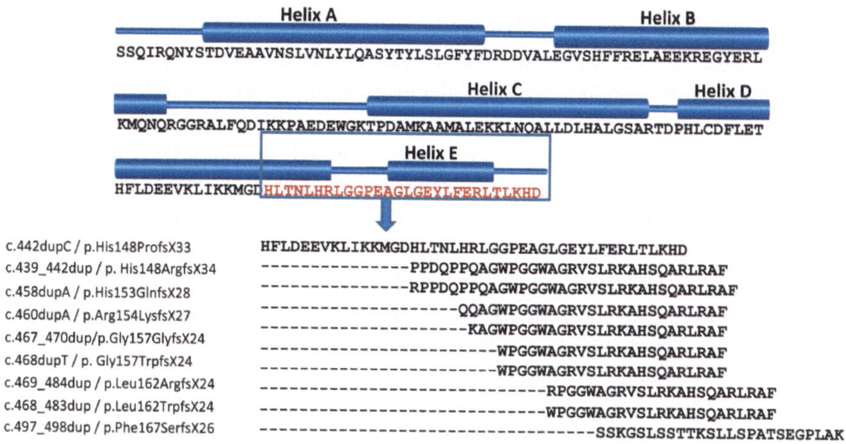

Fig. 9.1 Schematic representation of L-ferritin sequence and alignment of NF variants. Human L-ferritin sequence is matched with a schematic diagram of protein secondary structure elements that include five α-helices segments (blue cylinders), conventionally named with letters from A to E, connected by four loop elements and of two additional short amino acid sequences at the N and C termini. The blue rectangle indicates the sequence interval involved in the so far detected frameshift variants. Frameshift mutations are responsible for amino acidic changes at the C-terminus starting from different sequence position depending on the duplication starting point. In the lower panel, the alignment of the frameshift altered regions compared with the wild-type FtL sequence is shown. Dashes correspond to unchanged position, while modified residues are defined

Other cases may not yet have been identified; however, NF is a very rare disorder and thus the prevalence is not able to be calculated.

The c.460dupA mutation has been detected in the largest number of cases, 41 of which have been carefully described [18]. The mean age of onset was approximately 40 years, and the majority of patients presented with focal onset chorea or focal dystonia and reported a family history of movement disorders. The serum ferritin level was generally low. MR imaging was abnormal in all cases. The disease progressed relentlessly and became generalized in 5–10 years, leading to aphonia, dysphagia, and motor disability, with cognitive dysfunction as a late feature. Characteristic facial dystonia was common. Signs of parkinsonism, dementia, postural tremor, and cerebellar ataxia were also described in patients with other types of mutations [26, 86]. In general, the disorders appear to exhibit full penetrance at the age of 60 years, with heterogeneous clinical features associated with early movement disorders followed by minor cognitive impairments. The major and common feature is MRI signs, with similar abnormalities observed in all affected patients. In a comparison of the different NBIA disorders, NF was characterized by lesions in the globus pallidus, putamen, and dentate nuclei that differed from the other NBIA subtypes. Interestingly, two patients with NF showed the eye-of-the-tiger sign, which is considered pathognomonic almost exclusively for PKAN [78].

Brain histopathology has been performed on two patients with a c.460dupA mutation [22], one with a c.442dupC mutation [76, 91] and one with a c.497_498dup mutation [109]. Patients with c.460dupA mutations showed prominent abundant inclusions in the globus pallidus that contained ferritin and iron. Cysts formed in the basal ganglia centered on the site of the globus pallidus. Signs of neurodegeneration, such as axonal swelling and immunoreactivity for neurofilament, ubiquitin, and tau proteins, were observed throughout the white matter of the brain. Biochemical and hematological parameters were in the normal range, with the exception of low serum ferritin levels [22]. The patient with the c.442dupC mutation showed neurons and glia with highly distinctive swollen to vacuolated nuclei containing iron and ferritin [76]. Hyaline deposits were observed, which were positive for ferritin and iron. Moreover, the iron concentration in the putamen was >40-fold higher than in the control. The brain showed signs of oxidative damage and mitochondrial abnormalities. Further immunohistochemical staining of the specimen showed that the ferritin/iron deposits also contained neuroglobin [91], p53, and activated caspase-3, suggesting the activation of an apoptotic pathway. The patient with the c.497_498dup mutation presented with mild cerebral and cerebellar atrophy and cavitation of the putamen [109]. A major pathological alteration was the presence of intranuclear and intracytoplasmic bodies containing ferritin and iron in astrocytes and oligodendroglia in the gray and white matter. The ferritin bodies in the glia were more abundant in the caudate nucleus, putamen, and globus pallidus, where nerve loss was more severe. Similar bodies were observed in the fibroblasts of the papillary dermis, in the tubular epithelium of the kidney and in the endothelial cells of muscle capillaries.

According to the results from biochemical studies, the ferritin shells carrying the mutated subunits expressed from the c.460dupA and c.497_498dup alleles were less stable than wild-type ferritin and displayed largely different solubility and a lower capacity to incorporate iron [5, 6, 21]. Cellular models showed that the expression of these ferritin mutants not only increased the cellular free iron pool but also induced oxidative damage and impaired proteasome activity [20, 21]. Transgenic mice overexpressing the human c.497_498dup mutant showed movement disorders, the presence of cytosolic and nuclear inclusions typical of the human disease, and abnormalities in brain iron homeostasis [7, 110]. These data were further extended by a study of another c.497_498dup transgenic model that showed oxidative alterations in the brain [73]. Indeed, the examination of the postnatal hippocampal neurons obtained from this latter animal model revealed a major susceptibility to chronic iron overload and/or acute oxidative stress compared to wild-type neurons. Furthermore, the brain ultrastructural analyses highlighted an accumulation of lipofuscin granules associated with iron deposits that were particularly enriched in the cerebellum and striatum during aging [73]. Thus, oxidative stress and lipofuscin formation might be involved in the etiopathogenesis of human NF. The only data on human cells reported to date concerned fibroblasts obtained from a patient carrying the c.497_498dupTC mutation. These cells showed a consistent increase in the intracellular iron content, altered iron management, the accumulation of ferritin, and markers of oxidative stress, as detected both in patients and in the transgenic mice. ROS levels were significantly increased in these cells compared to controls [8].

Similar to aceruloplasminemia, the use of iron chelator (desferrioxamine, DFO, or deferiprone, DFP)-based therapies was tested both in animal models and in patients, obtaining similar results. A DFP treatment exerted remarkable effects on systemic iron homeostasis in mice with iron overload [36], while only marginal effects on iron deposition in the brain were observed [17], without signs of amelioration of the CNS pathology. In patients, iron chelator-based therapies cause substantial iron depletion without clinical benefits [18, 66].

The pathogenic process of NF hypothesized from the overall data predicts that ferritin/iron deposition triggers cellular oxidative damage and the subsequent cascade of oxidative events that lead to proteasomal and lysosomal dysfunction. Long-term impairments in the cellular recycling system permit the accumulation of lipofuscin in the brain tissue. The lysosome may undergo rupture that results in two effects: (i) direct cell damage by the hydrolytic enzymes and (ii) the release of lipofuscin in the cytosol, where it may function as a seed for further iron precipitation [73].

9.3 NBIA Related to CoA Biosynthesis

Two forms of NBIA are due to mutations in genes involved in the first and last steps of the same biosynthetic pathway that produces Coenzyme A (CoA). Mutations in these two genes, *PANK2* and *COASY*, cause the NBIA subtypes known as pantothenate kinase-associated neurodegeneration (PKAN, OMIM#234200) and CoA synthase protein-associated neurodegeneration (CoPAN, OMIM*609855), respectively.

PKAN approximately accounts for 50% of all NBIA cases and, depending on the age of onset and the rate of progression, is classified into two forms: the classical and the atypical forms. CoPAN appears to be more rare, as it has only been identified in five subjects to date [3, 31, 33]. These two disorders display strikingly similar phenotypes, presenting with early-onset spastic–dystonic paraparesis with a later appearance of parkinsonian features, cognitive impairment, obsessive–compulsive disorder, and brain iron accumulation [31]. PKAN usually manifests in early childhood as gait disturbances and rapidly progresses to a severe movement deficit with dystonia, dysarthria, and dysphagia [68]. The hallmark of this disease is the eye-of-the-tiger signal in the *globus pallidus* (*GP*) on T2-weighted magnetic resonance imaging, which reflects the focal accumulation of iron in this area [119]. The *substantia nigra* (*SN*) is also affected in patients with CoPAN and to a lesser extent in patients with PKAN. Iron accumulation correlates with neural damage and mitochondrial lesions and is often associated with axonal expansions (spheroids) but also appears as a granular form in perivascular sites [64].

Although iron accumulation is a hallmark of PKAN and CoPAN, its relationship with CoA dysfunctional biosynthesis remains unclear. CoA is a key molecule involved in more than 100 metabolic processes, among which CoA derivatives are crucial substrates for ATP generation mediated by the tricarboxylic acid cycle, fatty acid metabolism, cholesterol and ketone body biosynthesis, and histone and non-histone protein acetylation [1]. The CoA biosynthetic pathway involves five universal enzymatic steps that use pantothenate (vitamin B5), ATP, and cysteine. This pathway is initiated by PANK2, which converts pantothenic acid into 4′-phosphopantothenic acid. The last two steps of CoA biosynthesis are mediated by COASY. Although these processes are vital in any cell type, researchers have not yet explained why the disease primarily affects the central nervous system.

Several animal models have been developed in an attempt to clarify the molecular events that generate these disorders. *Pank2* null mice showed reduced growth, male infertility due to azoospermia [67], and impaired mitochondrial function [16], but did not suffer from movement disorders or exhibit signs of neurodegeneration, implying that the other *Pank* genes may compensate for the *Pank2* deficiency in mice. Only by modulating the fat content in the diet of these mice, it was possible to elicit neurodegeneration and cause a movement disorder associated with a general bioenergetic failure [15]. The *Drosophila Pank2* KO model (fumble) displays brain lesions and defective neurological functions, resulting in severe motor impairment [113]. The downregulation of *pank2* in zebra fish induces defects in neuronal development and in the formation of the vascular system [118].

A yeast model was used to study a human *COASY* pathogenic mutation. It showed mitochondrial dysfunction, and both the iron and lipid contents were altered [12]. Downregulation of zebra fish *coasy* expression led to a generalized reduction in size, poor definition of brain structures, and alteration of the vasculature [57]. Although all of these animal models were informative tools for studying pathological mechanisms, they share only a few neuropathological signs with milder severity than the human disorder, limiting their impact for predicting novel therapies [71]. In fact, while presenting some levels of neurodegeneration, these models lack any evidence of brain iron mishandling, preventing researchers from obtaining any insights into the causative link between CoA deficiency and brain iron deposition. Fibroblasts from patients with PKAN have been instrumental in revealing some defects in mitochondrial activity and iron metabolism associated with the *PANK2* deficiency [100]; however, their specific contributions to the pathological neurodegenerative processes have not been identified in these cells.

Insights into the neurodegenerative process began to emerge in an investigation of human neurons differentiated from induced pluripotent stem cells (iPSC) [87]. Compared with neurons from 3 healthy subjects, iPSC-derived glutamatergic neurons from 3 patients with PKAN showed alterations in mitochondrial function and impairments in mitochondrial iron-dependent biosynthesis that caused iron dyshomeostasis and subsequently increased ROS production, leading to major membrane excitability defects. In particular, the inability of the cells to maintain Fe/S biosynthesis simulates a cellular iron-deficient phenotype that promotes cellular iron uptake and increases the translation of TfR1 [87]. In addition, CoA supplementation prevented neuronal death and ROS formation by restoring mitochondrial and neuronal functions, suggesting that a CoA treatment is a possible therapeutic intervention [87]. A second paper analyzed iPSC-derived cortical neuronal cells obtained from fibroblasts from a patient with atypical PKAN [4]. Consistent with the milder phenotype, the Pank2 peptide was not abolished as described in the previous study but was present at lower levels than in the control cells. In this case, CoA homeostasis and cellular iron handling were normal; however, mitochondrial function was affected, with increased generation of reactive oxygen species and lipid peroxidation in patient-derived neurons [4]. Notably, an iron chelator treatment exacerbated the mitochondrial phenotype in neuronal cells from both control subjects and patients. Thus, both of these investigations using human neurons concluded that mitochondrial impairment is an early feature of the disease process and the subsequent ROS formation plays a main role in the pathogenesis; however, the iron chelator treatment did not appear to be a viable option. Indeed, the data on the efficacy of the DFP treatment collected from different clinical trials are still under investigation. Two pilot studies confirmed the safety and efficacy of the DFP treatment in reducing brain iron levels in patients with PKAN [19, 119], but the results from the still ongoing clinical trial involving approximately 100 patients with PKAN are needed to define the clinical outcomes of iron chelator treatments.

9.4 NBIA Caused by Defects in Proteins Related to Lipid Metabolism

Since sequencing has become a method for diagnosing NBIA, many similar cases in terms of symptomatology have been distinguished in different groups based on the genes in which the mutations were identified. The main group is composed of disorders associated with mutations in genes encoding proteins involved in lipid homeostasis. These diseases typically begin in childhood with symptoms such as developmental delay, regression or attention deficits associated with hyperactivity, spastic paraparesis, dystonia, and visual pathway disorders. Atypical variants begin in adolescence or adulthood, with a variable combination of dystonia, dysarthria, parkinsonism, and psychiatric symptoms. Iron accumulation is not a universal feature at the time when symptoms begin and is often documented only in the later stages, suggesting that iron overload, which is typically present in the *GP* and sometimes in the *SN*, is not the primary cause of the disease but contributes to neurodegeneration [29].

9.4.1 Neurodegeneration Associated with Phospholipase-A2G6 Defects (PLAN)

The first pathology to be distinguished from PKAN by sequencing was PLAN (or INAD, infantile neuroaxonal dystrophy, OMIM#256600), which is caused by defects in the PLA2G6 gene located on chromosome 22q13.1 that encodes a calcium-independent phospholipase involved in cell membrane homeostasis [116]. Defects in the protein may lead to altered lipid composition of the plasma, vesicular, and endosomal membranes, with subsequent alterations in membrane maintenance, causing neuronal damage [75]. This feature typically occurs in the first year of life as neuroaxonal childhood dystrophy with psychomotor regression and axial hypotonia that may be accompanied by neuropathy with spheroidal bodies in the peripheral nerves, which is detectable by biopsy. Brain MRI shows cerebellar atrophy in most patients, while only 40–50% of patients develop iron deposits. The atypical form manifests in the second or third decade of life as a form of dystonia-parkinsonism that is often accompanied by psychosis and dementia.

Studies on the *Pla2g6 -/-* mouse, which were also confirmed in a *Drosophila* model [58], suggested that one of the first pathological signs in the cells of the nervous system was the disruption of the crests of the inner mitochondrial membrane, followed by the destruction of the mitochondria, the release of cytochrome c, and the formation of swollen and degenerated axons as the disease progresses [75]. Even the presynaptic membranes were altered in this animal model, indicating that the mitochondria require the remodeling activity of this phospholipase [10]. A more recent study of the *Pla2g6 -/-* mouse model revealed that these mice showed age-dependent iron accumulation in the *SN*, *ST,* and *GP* regions of the brain, similar to findings in affected patients. This iron deposition is associated with increased levels of iron-related proteins such as DMT1, TfR1, and IRPs, decreased FPN1 levels, enhanced lipid peroxidation, and mitochondrial dysfunction [9]. These features are observed before signs of neuronal loss, suggesting that iron-dependent ROS formation might be a primary cause of neurodegeneration [9]. Evidence of an impairment in the recycling of transferrin and transferrin receptor from endosomes due to the loss of *PLA2G6* was also reported in vitro in HeLa cells [23], further highlighting the connection between the iron imbalance and impaired dynamics of cellular membranes.

9.4.2 Mitochondrial Membrane Protein-Associated Neurodegeneration (MPAM)

MPAM accounts for approximately 30% of the NBIA cases. It is caused by a mutation in *C19orf12* (OMIM*614297) that encodes a mitochondrial protein with an unknown function but is probably involved in fatty acid metabolism and in the degradation of the branched amino acid chains [46, 88]. The main symptoms of MPAN are progressive spastic para/tetraparesis, generalized dystonia, motor axonal neuropathy, parkinsonism, psychiatric symptoms, retinal abnormalities, and optic atrophy [46, 50, 70]. Patients with MPAN develop Lewy bodies containing α-synuclein in the *GP* and midbrain, as well as perivascular iron deposits similar to those observed in patients with PKAN [49]. The *C19orf12*-encoded peptide is a transmembrane glycine zipper-containing protein with two isoforms originating from different start codons. In the literature, few reports have characterized the protein. An analysis of HeLa subcellular fractions indicated that the longer isoform was located in the mitochondria, endoplasmic reticulum, and a small fraction in the MAM [107], suggesting a role in the transport of phospholipids. Fibroblasts from one patient showed an alteration of mitochondrial calcium homeostasis and increased H_2O_2-induced apoptosis compared to controls [107]. The only available animal disease model is the knockdown of the C19orf12 orthologues in *Drosophila* that showed signs of neurological defects and the presence of vacuoles in the brain and optical lobe, but not iron accumulation [53].

9.4.3 Fatty Acid Hydroxylase-Associated Neurodegeneration (FAHN)

FAHN (OMIM#612319) is caused by mutations in the gene encoding FA2H (fatty acid-2-hydroxylase). The disorder was initially identified in 2 consanguineous families, one Italian and one Albanian [65]. To date, 51 patients have been reported to carry missense mutations [77], some of whom were previously diagnosed with leukodystrophy with spastic paraparesis and dystonia and a form of paraplegia spasticity [101]. The pathology is initially characterized by spasticity, ataxia, dystonia, optic atrophy, and oculomotor abnormalities, followed by cognitive disorders and epilepsy. Magnetic resonance studies in patients have shown iron overload at the level of the *GP*, *SN*, subcortical, and periventricular regions [40], as well as profound changes in the white matter distribution, cerebellar atrophy, and a thinning of the *corpus callosum*. In the mouse model, an accumulation of microglial cells was observed in areas exhibiting greater degeneration of white matter [90]. FA2H is a membrane-bound protein involved in the synthesis of ceramide, an essential component of myelin, and its deficit caused defects in myelination and a subsequent impairment in axonal function [90].

To date, no specific analysis of iron metabolism has been reported using models of this disorder. The evidence that the protein includes a cytochrome b5-like domain, a heme group, and a catalytic di-iron cluster involved in the enzymatic activity of the protein suggests a possible direct interaction with other iron-containing proteins. A recent computational analysis of the effect of the causative missense mutations on the 3D protein structure indicated that these mutations interfere with the heme binding affinity, affecting protein stability, or alter the structure of the catalytic hydroxylase domain, disturbing enzymatic activity [77]. However, researchers have not determined how the impaired enzymatic activity leads to brain iron accumulation; we can only state that the impairment in myelin synthesis is associated with alterations in brain iron homeostasis.

9.4.4 Leukoencephalopathy with Dystonia and Motor Neuropathy

Leukoencephalopathy with dystonia and motor neuropathy (LKDMN, OMIM# 613724) is an autosomal recessive disorder caused by *SCP2* mutations that was only recently classified as an NBIA disorder. It is characterized by azoospermia, tremor, slight cerebellar ataxia, and balance and gait impairments. MRI shows bilateral hyperintense signals in the thalamus, butterfly-like lesions in the pons, and lesions in the occipital region [34, 51]. The sterol carrier protein-2 (SCP2) gene encodes two different proteins located in peroxisomes: sterol carrier protein X (SCPx) and sterol carrier protein 2 (SCP2). These proteins possess the thiolase activity required to degrade branched-chain fatty acids [111]. Patients carrying *SCP2* mutations show abnormal fatty acid acyl-CoA metabolism, suggesting a common pathogenic mechanism between this disorder and the NBIA related to CoA biosynthesis.

9.4.5 **CRAT**- *and* **REPS1**-*Associated Neurodegeneration*

Two other genes involved in lipid homeostasis have recently been identified as causative for NBIA [28]. The genes are *CRAT* (OMIM*600184), which is involved in the transfer of acyl groups from carnitine to CoA and in the transport of fatty acids for beta-oxidation, and *REPS1* (OMIM*614825), which is engaged in endocytosis and vesicle transport. Interestingly, the authors also described a new perspective of the NBIA pathogenic mechanism. They observed a common phenotype in fibroblasts derived from patients with *CRAT* and *REPS1* mutations, as well as fibroblasts derived from patients with *PANK2*, *PLA2G6*, *C19orf12*, and *FA2H* mutations. All these cells exhibited reduced TfR1 palmitoylation that affected the recycling of the receptor [28]. This anomaly in cellular iron incorporation led to iron overload that was decreased by treatment with the antimalarial drug artesunate. This drug promotes receptor palmitoylation, restoring the normal TfR1 recycling [28]. This report was the first time to experimentally confirm a common pathogenic mechanism for NBIA disorders.

9.5 NBIA Related to Lysosome Regulation/Autophagosome Formation

Although the role of lysosomes in iron metabolism is known [69], some lysosomal disorders have only recently been included in the NBIA categories. These disorders are caused by mutations in genes encoding lysosomal proteins and show a variable age of onset and severity of the phenotype, depending on the degree of dysfunctionality of the mutated proteins. An explanation for why only a portion of the affected patients exhibit iron overload in the brain is unknown, and researchers have proposed that iron is not the primary cause of the disease in this case [29].

9.5.1 Kufor–Rakeb Disease (KRD)

KRD (OMIM#606693) was originally described in a Jordanian family, but the causative gene was identified only later in a large group of Chilean brothers [11]. Then, other cases and new mutations have been identified in several countries, including Brazil, Pakistan, Afghanistan, Japan, and Italy [101]. The gene associated with this pathology is the *ATP13A2* gene [94] located on chromosome 1p36.13. It contains 29 exons and encodes a P-type ATPase. In the human brain, this protein is expressed in the pyramidal neurons of the cerebral cortex and the dopaminergic neurons of the *SN* [95]. KRD usually manifests as a juvenile form of parkinsonism (also classified as PARK9) that is responsive to levodopa. Patients show pyramidal symptoms, cognitive decline, and visual hallucinations; patients with the late-onset form may also exhibit ataxia and axonal neuropathy [32]. The majority of patients do not present with iron overload; yet, when it is present, it is mainly found in the *striatum* region.

The peptide encoded by *ATP13A2* is located in lysosomes and late endosomes and was proposed to remodel the endocytic pathway toward export [25]. The ablation of this gene in mice led to lysosomal dysfunction, lipofuscin accumulation, α-synuclein aggregation, and age-dependent sensorimotor deficits [102]. Recently, the *ATP13A2* gene was identified as a novel *HIF1α* target and therefore regulated by the PHD2-HIF1α signaling pathway. Indeed, *ATP13A2* knockdown abrogated the maintenance of cellular iron homeostasis elicited by PHD2 inhibition, which occurred under conditions of mitochondrial stress both in vivo and in cultured dopaminergic cells [93]. The authors concluded that *ATP13A2* is involved in maintaining lysosomal iron stores.

9.5.2 β-Propeller Protein-Associated Neurodegeneration (BPAM)

β-Propeller-associated neurodegeneration (BPAN) (OMIM#300894) is caused by mutations in the WD repeat domain 45 (*WDR45*) gene located on the X chromosome. To date, all affected individuals are the only ones in their family with the disease and most are females. The causative mutations are de novo mutations and are lethal to most male individuals [39]. Indeed, patients of both genders displayed similar symptoms due to the presence of somatic mosaicism in male patients and germline or somatic mutations in females, as well as skewing of X chromosome inactivation [41]. Affected individuals exhibited a developmental delay during infancy while developing progressive parkinsonism associated with dementia during adolescence and adulthood [41]. Magnetic resonance studies have revealed iron accumulation in the *GP* and *SN*. In particular, T1-weighted images revealed a hypo-intense zone surrounded by a hyperintense area within the *SN*.

The *WDR45* gene encodes a protein belonging to the WD40 scaffold-protein family that is involved in various cellular processes, such as cell cycle progression, signal transduction, apoptosis, and regulation of gene expression. In particular, Atg18, the yeast orthologue of the human *WDR45* gene, is a crucial component of the complex required for autophagosome formation [92]. Recent reports have highlighted the role of iron in the pathogenesis of the BPAN disorder. Studies on fibroblasts from two individuals with BPAN revealed that an upregulation of the ferrous iron transport (-)IRE/DMT1 and downregulation of TfR1 compared with control subjects. This alteration in the levels of iron transporter proteins occurs concomitantly with the increase in intracellular ferrous iron levels under starvation conditions [52], suggesting that the compromised autophagy in BPAN cells may determine iron overload [52]. A further study of *WDR45* mutant fibroblasts and iPSC-derived midbrain neurons confirmed and extended these observations. The authors confirmed that the loss of *WDR45* increased cellular iron levels and promoted oxidative stress, accompanied by mitochondrial abnormalities, autophagy defects, and diminished lysosomal function [104]. Therefore, the close relationship between the diminished lysosomal function and the onset of iron accumulation has also been highlighted in these disorders.

9.5.3 AP4M1-*Associated Neurodegeneration*

A new candidate NBIA disorder was reported in a recent study that identified iron deposition in the brains of 3 patients afflicted with an autosomal recessive form of hereditary spastic paraplegia (HSP) (OMIM#602296) [98] that is characterized by early-onset developmental delay, tetraparesis, motor function deterioration, and intellectual disability. This very rare neurodevelopmental disorder is due to mutations in *AP4M1*. The *AP4M1* gene encodes the subunits of adaptor protein 4 (AP-4), a heterotetrameric complex involved in intracellular protein trafficking [24]. No data on iron involvement in the pathophysiology of the disorder have been reported; however, the finding that fibroblasts from AP-4-deficient patients and neurons from AP-4-deficient mice show impaired autophagy further emphasizes the stringent link between alterations in autophagy and iron deposition.

9.6 NBIA Related to Genes with Unknown Functions

Two other disorders are classified as NBIA, although not all patients show signs of brain iron deposition. These disorders are briefly described here because published data concerning the role of iron in the pathological process do not exist.

9.6.1 Woodhouse–Sakati Syndrome

Woodhouse–Sakati syndrome (OMIM#241080) is a rare autosomal condition caused by mutations in the *DCAF17* gene (also known as *C2orf37*), which encodes a nucleolar protein with an as-yet-unknown function. The disorder manifests as hypogonadism, deafness, alopecia, and diabetes mellitus; subsequently, with the progression of the disease, dysarthria, dystonia, chorea, and cognitive decline are observed. Magnetic resonance studies revealed lesions in the white matter and iron deposits in *SN* and *GP* in a portion of affected patients [2].

9.6.2 GTPBP2*-Associated Neurodegeneration*

A splice site mutation in *GTPBP2* (OMIM*607434) was identified in 3 affected siblings in a consanguineous family. The patients exhibited mental retardation, ataxia, and dystonic features [54]. Iron accumulation in the brains of the patients was a notable phenotypic trait. However, recently, biallelic inactivating variants of *GTPBP2* were reported in three unrelated families with a neurological phenotype that differed from the previously published family and did not display signs of iron accumulation [13]. The *GTPBP2* gene encodes a GTP-binding protein that is a member of the GTPase superfamily that is capable of binding GTP or GDP. The precise function of *GTPBP2* is not yet defined, but it has been proposed to be a regulator of BMP and Wnt signaling [37] and to be involved in the dysregulation of mRNA translation [117].

9.7 Conclusions

Advances in genetic screens based on whole-exome sequencing and the recent insights into the biochemical basis of NBIA have begun to delineate the role of iron in the pathophysiology of these disorders. In fact, even if a genetic cause has not been defined in all cases of NBIA, the recent identification of new genes has assisted researchers in outlining some crucial interrelated cellular pathways that, when impaired, induce iron accumulation in the CNS. The pathways encompass mitochondrial functions, lipid metabolism, membrane remodeling, and autophagy. Alterations in all these pathways result in the dysregulation of iron management. Mitochondrial function is essential for the transformation of iron in its biological form (i.e., iron–sulfur cluster and heme cofactors) and for the preservation of energy production processes that depend on the availability of the same iron cofactors. Indeed, the levels of these cofactors are also decreased in patients with a CoA deficiency, further highlighting the relationship between iron availability, energy resources, and lipid homeostasis. Lipid metabolism also plays a crucial role in myelin perturbation, where iron is required for its biosynthesis [97]. Membrane remodeling and autophagy appear to be essential to avoid iron deposition in the brain. In particular, the impaired ability of cells to disrupt the iron-containing macromolecules (i.e., ferritin) and organelles (mitochondria) through the lysosomal pathway appears to be a common phenotype in patients with various NBIAs.

Although additional in vitro and in vivo studies are needed to define whether iron dysregulation has a primary or an auxiliary role in the pathophysiology of these diseases, the clarification of the pathogenic mechanisms of NBIAs will add important knowledge required for a better understanding of the relationship between iron and the neurodegenerative process that is necessary to define the therapeutic options.

Acknowledgements The financial support from Telethon-Italia (Grant nos. GGP10099, GGP11088 and GGP16234 to SL) and AISNAF (to SL) is gratefully acknowledged. The authors are grateful to Dr. Ermanna Rovida for assisting with the figure preparation.

References

1. Akram M (2014) Citric acid cycle and role of its intermediates in metabolism. Cell Biochem Biophys 68:475–478
2. Alazami AM, Al-Saif A, Al-Semari A, Bohlega S, Zlitni S, Alzahrani F, Bavi P, Kaya N, Colak D, Khalak H, Baltus A, Peterlin B, Danda S, Bhatia KP, Schneider SA, Sakati N, Walsh CA, Al-Mohanna F, Meyer B, Alkuraya FS (2008) Mutations in C2orf37, encoding a nucleolar protein, cause hypogonadism, alopecia, diabetes mellitus, mental retardation, and extrapyramidal syndrome. Am J Hum Genet 83:684–691
3. Annesi G, Gagliardi M, Iannello G, Quattrone A (2016) Mutational analysis of COASY in an Italian patient with NBIA. Parkinsonism Relat Disord 28:150–151
4. Arber C, Angelova PR, Wiethoff S, Tsuchiya Y, Mazzacuva F, Preza E, Bhatia KP, Mills K, Gout I, Abramov AY, Hardy J, Duce JA, Houlden H, Wray S (2017) iPSC-derived neuronal models of PANK2-associated neurodegeneration reveal mitochondrial dysfunction contributing to early disease. PLoS ONE 12:e0184104
5. Baraibar MA, Barbeito AG, Muhoberac BB, Vidal R (2008) Iron-mediated aggregation and a localized structural change characterize ferritin from a mutant light chain polypeptide that causes neurodegeneration. J Biol Chem 283:31679–31689
6. Baraibar MA, Muhoberac BB, Garringer HJ, Hurley TD, Vidal R (2010) Unraveling of the E-helices and disruption of 4-fold pores are associated with iron mishandling in a mutant ferritin causing neurodegeneration. J Biol Chem 285:1950–1956
7. Barbeito AG, Garringer HJ, Baraibar MA, Gao X, Arredondo M, Nunez MT, Smith MA, Ghetti B, Vidal R (2009) Abnormal iron metabolism and oxidative stress in mice expressing a mutant form of the ferritin light polypeptide gene. J Neurochem 109:1067–1078
8. Barbeito AG, Levade T, Delisle MB, Ghetti B, Vidal R (2010) Abnormal iron metabolism in fibroblasts from a patient with the neurodegenerative disease hereditary ferritinopathy. Mol Neurodegener 5:50
9. Beck G, Shinzawa K, Hayakawa H, Baba K, Yasuda T, Sumi-Akamaru H, Tsujimoto Y, Mochizuki H (2015) Deficiency of calcium-independent phospholipase A2 Beta induces brain iron accumulation through upregulation of divalent metal transporter 1. PLoS ONE 10:e0141629
10. Beck G, Sugiura Y, Shinzawa K, Kato S, Setou M, Tsujimoto Y, Sakoda S, Sumi-Akamaru H (2011) Neuroaxonal dystrophy in calcium-independent phospholipase A2beta deficiency

results from insufficient remodeling and degeneration of mitochondrial and presynaptic membranes. J Neurosci 31:11411–11420

11. Behrens MI, Bruggemann N, Chana P, Venegas P, Kagi M, Parrao T, Orellana P, Garrido C, Rojas CV, Hauke J, Hahnen E, Gonzalez R, Seleme N, Fernandez V, Schmidt A, Binkofski F, Kompf D, Kubisch C, Hagenah J, Klein C, Ramirez A (2010) Clinical spectrum of Kufor-Rakeb syndrome in the Chilean kindred with ATP13A2 mutations. Mov Disord 25:1929–1937
12. Berti CC, Dallabona C, Lazzaretti M, Dusi S, Tosi E, Tiranti V, Goffrini P (2015) Modeling human Coenzyme A synthase mutation in yeast reveals altered mitochondrial function, lipid content and iron metabolism. Microb Cell 2:126–135
13. Bertoli-Avella AM, Garcia-Aznar JM, Brandau O, Al-Hakami F, Yuksel Z, Marais A, Gruning NM, Abbasi Moheb L, Paknia O, Alshaikh N, Alameer S, Marafi MJ, Al-Mulla F, Al-Sannaa N, Rolfs A, Bauer P (2018) Biallelic inactivating variants in the GTPBP2 gene cause a neurodevelopmental disorder with severe intellectual disability. Eur J Hum Genet 26:592–598
14. Brissot P, Ropert M, Le Lan C, Loreal O (2012) Non-transferrin bound iron: a key role in iron overload and iron toxicity. Biochim Biophys Acta 1820:403–410
15. Brunetti D, Dusi S, Giordano C, Lamperti C, Morbin M, Fugnanesi V, Marchet S, Fagiolari G, Sibon O, Moggio M, d'Amati G, Tiranti V (2014) Pantethine treatment is effective in recovering the disease phenotype induced by ketogenic diet in a pantothenate kinase-associated neurodegeneration mouse model. Brain 137:57–68
16. Brunetti D, Dusi S, Morbin M, Uggetti A, Moda F, D'Amato I, Giordano C, d'Amati G, Cozzi A, Levi S, Hayflick S, Tiranti V (2012) Pantothenate kinase-associated neurodegeneration: altered mitochondria membrane potential and defective respiration in Pank2 knock-out mouse model. Hum Mol Genet 21:5294–5305
17. Capoccia S, Maccarinelli F, Buffoli B, Rodella LF, Cremona O, Arosio P, Cirulli F (2015) Behavioral characterization of mouse models of neuroferritinopathy. PLoS ONE 10:e0118990
18. Chinnery PF, Crompton DE, Birchall D, Jackson MJ, Coulthard A, Lombes A, Quinn N, Wills A, Fletcher N, Mottershead JP, Cooper P, Kellett M, Bates D, Burn J (2007) Clinical features and natural history of neuroferritinopathy caused by the FTL1 460InsA mutation. Brain 130:110–119
19. Cossu G, Abbruzzese G, Matta G, Murgia D, Melis M, Ricchi V, Galanello R, Barella S, Origa R, Balocco M, Pelosin E, Marchese R, Ruffinengo U, Forni GL (2014) Efficacy and safety of deferiprone for the treatment of pantothenate kinase-associated neurodegeneration (PKAN) and neurodegeneration with brain iron accumulation (NBIA): results from a four years follow-up. Parkinsonism Relat Disord 20:651–654
20. Cozzi A, Rovelli E, Frizzale G, Campanella A, Amendola M, Arosio P, Levi S (2010) Oxidative stress and cell death in cells expressing L-ferritin variants causing neuroferritinopathy. Neurobiol Dis 37:77–85
21. Cozzi A, Santambrogio P, Corsi B, Campanella A, Arosio P, Levi S (2006) Characterization of the l-ferritin variant 460InsA responsible of a hereditary ferritinopathy disorder. Neurobiol Dis 23:644–652
22. Curtis AR, Fey C, Morris CM, Bindoff LA, Ince PG, Chinnery PF, Coulthard A, Jackson MJ, Jackson AP, McHale DP, Hay D, Barker WA, Markham AF, Bates D, Curtis A, Burn J (2001) Mutation in the gene encoding ferritin light polypeptide causes dominant adult-onset basal ganglia disease. Nat Genet 28:350–354
23. de Figueiredo P, Doody A, Polizotto RS, Drecktrah D, Wood S, Banta M, Strang MS, Brown WJ (2001) Inhibition of transferrin recycling and endosome tubulation by phospholipase A2 antagonists. J Biol Chem 276:47361–47370
24. De Pace R, Skirzewski M, Damme M, Mattera R, Mercurio J, Foster AM, Cuitino L, Jarnik M, Hoffmann V, Morris HD, Han TU, Mancini GMS, Buonanno A, Bonifacino JS (2018) Altered distribution of ATG9A and accumulation of axonal aggregates in neurons from a mouse model of AP-4 deficiency syndrome. PLoS Genet 14:e1007363
25. de Tezanos Pinto F, Adamo HP (2018) The strategic function of the P5-ATPase ATP13A2 in toxic waste disposal. Neurochem Int 112:108–113

26. Devos D, Tchofo PJ, Vuillaume I, Destee A, Batey S, Burn J, Chinnery PF (2009) Clinical features and natural history of neuroferritinopathy caused by the 458dupA FTL mutation. Brain 132:e109
27. di Patti MC, Maio N, Rizzo G, De Francesco G, Persichini T, Colasanti M, Polticelli F, Musci G (2009) Dominant mutants of ceruloplasmin impair the copper loading machinery in aceruloplasminemia. J Biol Chem 284:4545–4554
28. Drecourt A, Babdor J, Dussiot M, Petit F, Goudin N, Garfa-Traore M, Habarou F, Bole-Feysot C, Nitschke P, Ottolenghi C, Metodiev MD, Serre V, Desguerre I, Boddaert N, Hermine O, Munnich A, Rotig A (2018) Impaired transferrin receptor palmitoylation and recycling in neurodegeneration with brain iron accumulation. Am J Hum Genet 102:266–277
29. Dusek P, Schneider SA (2012) Neurodegeneration with brain iron accumulation. Curr Opin Neurol 25:499–506
30. Dusek P, Schneider SA (2013) Neurodegenerative disorder with brain iron accumulation previously known as SENDA syndrome now genetically determined. Mov Disord 28:1051–1052
31. Dusi S, Valletta L, Haack TB, Tsuchiya Y, Venco P, Pasqualato S, Goffrini P, Tigano M, Demchenko N, Wieland T, Schwarzmayr T, Strom TM, Invernizzi F, Garavaglia B, Gregory A, Sanford L, Hamada J, Bettencourt C, Houlden H, Chiapparini L, Zorzi G, Kurian MA, Nardocci N, Prokisch H, Hayflick S, Gout I, Tiranti V (2014) Exome sequence reveals mutations in CoA synthase as a cause of neurodegeneration with brain iron accumulation. Am J Hum Genet 94:11–22
32. Eiberg H, Hansen L, Korbo L, Nielsen IM, Svenstrup K, Bech S, Pinborg LH, Friberg L, Hjermind LE, Olsen OR, Nielsen JE (2012) Novel mutation in ATP13A2 widens the spectrum of Kufor-Rakeb syndrome (PARK9). Clin Genet 82:256–263
33. Evers C, Seitz A, Assmann B, Opladen T, Karch S, Hinderhofer K, Granzow M, Paramasivam N, Eils R, Diessl N, Bartram CR, Moog U (2017) Diagnosis of CoPAN by whole exome sequencing: waking up a sleeping tiger's eye. Am J Med Genet A
34. Ferdinandusse S, Kostopoulos P, Denis S, Rusch H, Overmars H, Dillmann U, Reith W, Haas D, Wanders RJ, Duran M, Marziniak M (2006) Mutations in the gene encoding peroxisomal sterol carrier protein X (SCPx) cause leukencephalopathy with dystonia and motor neuropathy. Am J Hum Genet 78:1046–1052
35. Finkenstedt A, Wolf E, Hofner E, Gasser BI, Bosch S, Bakry R, Creus M, Kremser C, Schocke M, Theurl M, Moser P, Schranz M, Bonn G, Poewe W, Vogel W, Janecke AR, Zoller H (2010) Hepatic but not brain iron is rapidly chelated by deferasirox in aceruloplasminemia due to a novel gene mutation. J Hepatol 53:1101–1107
36. Garringer HJ, Irimia JM, Li W, Goodwin CB, Richine B, Acton A, Chan RJ, Peacock M, Muhoberac BB, Ghetti B, Vidal R (2016) Effect of systemic iron overload and a chelation therapy in a mouse model of the neurodegenerative disease hereditary Ferritinopathy. PLoS ONE 11:e0161341
37. Gillis WQ, Kirmizitas A, Iwasaki Y, Ki DH, Wyrick JM, Thomsen GH (2016) Gtpbp2 is a positive regulator of Wnt signaling and maintains low levels of the Wnt negative regulator Axin. Cell Commun Signal 14:15
38. Gonzalez-Cuyar LF, Perry G, Miyajima H, Atwood CS, Riveros-Angel M, Lyons PF, Siedlak SL, Smith MA, Castellani RJ (2008) Redox active iron accumulation in aceruloplasminemia. Neuropathology 28:466–471
39. Gregory A, Hayflick SJ (2011) Genetics of neurodegeneration with brain iron accumulation. Curr Neurol Neurosci Rep 11:254–261
40. Gregory A, Venkateswaran S, Hayflick SJ (2011) Fatty acid hydroxylase-associated neurodegeneration. In: Adam MP, Ardinger HH, Pagon RA, Wallace SE, Bean LJH, Stephens K, Amemiya A (eds) GeneReviews((R)). Seattle, pp 1993–2018
41. Haack TB, Hogarth P, Kruer MC, Gregory A, Wieland T, Schwarzmayr T, Graf E, Sanford L, Meyer E, Kara E, Cuno SM, Harik SI, Dandu VH, Nardocci N, Zorzi G, Dunaway T, Tarnopolsky M, Skinner S, Frucht S, Hanspal E, Schrander-Stumpel C, Heron D, Mignot C, Garavaglia B, Bhatia K, Hardy J, Strom TM, Boddaert N, Houlden HH, Kurian MA, Meitinger T, Prokisch H, Hayflick SJ (2012) Exome sequencing reveals de novo WDR45

mutations causing a phenotypically distinct, X-linked dominant form of NBIA. Am J Hum Genet 91:1144–1149

42. Hadziahmetovic M, Dentchev T, Song Y, Haddad N, He X, Hahn P, Pratico D, Wen R, Harris ZL, Lambris JD, Beard J, Dunaief JL (2008) Ceruloplasmin/hephaestin knockout mice model morphologic and molecular features of AMD. Invest Ophthalmol Vis Sci 49:2728–2736
43. Hadziahmetovic M, Song Y, Wolkow N, Iacovelli J, Grieco S, Lee J, Lyubarsky A, Pratico D, Connelly J, Spino M, Harris ZL, Dunaief JL (2011) The oral iron chelator deferiprone protects against iron overload-induced retinal degeneration. Invest Ophthalmol Vis Sci 52:959–968
44. Hahn P, Qian Y, Dentchev T, Chen L, Beard J, Harris ZL, Dunaief JL (2004) Disruption of ceruloplasmin and hephaestin in mice causes retinal iron overload and retinal degeneration with features of age-related macular degeneration. Proc Natl Acad Sci U S A 101:13850–13855
45. Harris ZL, Durley AP, Man TK, Gitlin JD (1999) Targeted gene disruption reveals an essential role for ceruloplasmin in cellular iron efflux. Proc Natl Acad Sci U S A 96:10812–10817
46. Hartig MB, Iuso A, Haack T, Kmiec T, Jurkiewicz E, Heim K, Roeber S, Tarabin V, Dusi S, Krajewska-Walasek M, Jozwiak S, Hempel M, Winkelmann J, Elstner M, Oexle K, Klopstock T, Mueller-Felber W, Gasser T, Trenkwalder C, Tiranti V, Kretzschmar H, Schmitz G, Strom TM, Meitinger T, Prokisch H (2011) Absence of an orphan mitochondrial protein, c19orf12, causes a distinct clinical subtype of neurodegeneration with brain iron accumulation. Am J Hum Genet 89:543–550
47. Hayflick SJ, Kurian MA, Hogarth P (2018) Neurodegeneration with brain iron accumulation. Handb Clin Neurol 147:293–305
48. Hellman NE, Kono S, Mancini GM, Hoogeboom AJ, De Jong GJ, Gitlin JD (2002) Mechanisms of copper incorporation into human ceruloplasmin. J Biol Chem 277:46632–46638
49. Hogarth P, Gregory A, Kruer MC, Sanford L, Wagoner W, Natowicz MR, Egel RT, Subramony SH, Goldman JG, Berry-Kravis E, Foulds NC, Hammans SR, Desguerre I, Rodriguez D, Wilson C, Diedrich A, Green S, Tran H, Reese L, Woltjer RL, Hayflick SJ (2013) New NBIA subtype: genetic, clinical, pathologic, and radiographic features of MPAN. Neurology 80:268–275
50. Horvath R, Holinski-Feder E, Neeve VC, Pyle A, Griffin H, Ashok D, Foley C, Hudson G, Rautenstrauss B, Nurnberg G, Nurnberg P, Kortler J, Neitzel B, Bassmann I, Rahman T, Keavney B, Loughlin J, Hambleton S, Schoser B, Lochmuller H, Santibanez-Koref M, Chinnery PF (2012) A new phenotype of brain iron accumulation with dystonia, optic atrophy, and peripheral neuropathy. Mov Disord 27:789–793
51. Horvath R, Lewis-Smith D, Douroudis K, Duff J, Keogh M, Pyle A, Fletcher N, Chinnery PF (2015) SCP2 mutations and neurodegeneration with brain iron accumulation. Neurology 85:1909–1911
52. Ingrassia R, Memo M, Garavaglia B (2017) Ferrous iron up-regulation in fibroblasts of patients with beta propeller protein-associated neurodegeneration (BPAN). Front Genet 8:18
53. Iuso A, Sibon OC, Gorza M, Heim K, Organisti C, Meitinger T, Prokisch H (2014) Impairment of Drosophila orthologs of the human orphan protein C19orf12 induces bang sensitivity and neurodegeneration. PLoS ONE 9:e89439
54. Jaberi E, Rohani M, Shahidi GA, Nafissi S, Arefian E, Soleimani M, Rasooli P, Ahmadieh H, Daftarian N, KaramiNejadRanjbar M, Klotzle B, Fan JB, Turk C, Steemers F, Elahi E (2016) Identification of mutation in GTPBP2 in patients of a family with neurodegeneration accompanied by iron deposition in the brain. Neurobiol Aging 38:216 e211–216 e218
55. Jeong SY, David S (2003) Glycosylphosphatidylinositol-anchored ceruloplasmin is required for iron efflux from cells in the central nervous system. J Biol Chem 278:27144–27148
56. Kaneko K, Hineno A, Yoshida K, Ohara S, Morita H, Ikeda S (2012) Extensive brain pathology in a patient with aceruloplasminemia with a prolonged duration of illness. Hum Pathol 43:451–456
57. Khatri D, Zizioli D, Tiso N, Facchinello N, Vezzoli S, Gianoncelli A, Memo M, Monti E, Borsani G, Finazzi D (2016) Down-regulation of coasy, the gene associated with NBIA-VI, reduces Bmp signaling, perturbs dorso-ventral patterning and alters neuronal development in zebrafish. Sci Rep 6:37660

58. Kinghorn KJ, Castillo-Quan JI, Bartolome F, Angelova PR, Li L, Pope S, Cocheme HM, Khan S, Asghari S, Bhatia KP, Hardy J, Abramov AY, Partridge L (2015) Loss of PLA2G6 leads to elevated mitochondrial lipid peroxidation and mitochondrial dysfunction. Brain 138:1801–1816
59. Kono S (2012) Aceruloplasminemia. Curr Drug Targets 13:1190–1199
60. Kono S (2013) Aceruloplasminemia: an update. Int Rev Neurobiol 110:125–151
61. Kono S, Miyajima H (2006) Molecular and pathological basis of aceruloplasminemia. Biol Res 39:15–23
62. Kono S, Suzuki H, Oda T, Shirakawa K, Takahashi Y, Kitagawa M, Miyajima H (2007) Cys-881 is essential for the trafficking and secretion of truncated mutant ceruloplasmin in aceruloplasminemia. J Hepatol 47:844–850
63. Kono S, Yoshida K, Tomosugi N, Terada T, Hamaya Y, Kanaoka S, Miyajima H (2010) Biological effects of mutant ceruloplasmin on hepcidin-mediated internalization of ferroportin. Biochim Biophys Acta 1802:968–975
64. Kruer MC, Hiken M, Gregory A, Malandrini A, Clark D, Hogarth P, Grafe M, Hayflick SJ, Woltjer RL (2011) Novel histopathologic findings in molecularly-confirmed pantothenate kinase-associated neurodegeneration. Brain 134:947–958
65. Kruer MC, Paisan-Ruiz C, Boddaert N, Yoon MY, Hama H, Gregory A, Malandrini A, Woltjer RL, Munnich A, Gobin S, Polster BJ, Palmeri S, Edvardson S, Hardy J, Houlden H, Hayflick SJ (2010) Defective FA2H leads to a novel form of neurodegeneration with brain iron accumulation (NBIA). Ann Neurol 68:611–618
66. Kubota A, Hida A, Ichikawa Y, Momose Y, Goto J, Igeta Y, Hashida H, Yoshida K, Ikeda S, Kanazawa I, Tsuji S (2009) A novel ferritin light chain gene mutation in a Japanese family with neuroferritinopathy: description of clinical features and implications for genotype-phenotype correlations. Mov Disord 24:441–445
67. Kuo YM, Duncan JL, Westaway SK, Yang H, Nune G, Xu EY, Hayflick SJ, Gitschier J (2005) Deficiency of pantothenate kinase 2 (Pank2) in mice leads to retinal degeneration and azoospermia. Hum Mol Genet 14:49–57
68. Kurian MA, Hayflick SJ (2013) Pantothenate kinase-associated neurodegeneration (PKAN) and PLA2G6-associated neurodegeneration (PLAN): review of two major neurodegeneration with brain iron accumulation (NBIA) phenotypes. Int Rev Neurobiol 110:49–71
69. Kurz T, Eaton JW, Brunk UT (2011) The role of lysosomes in iron metabolism and recycling. Int J Biochem Cell Biol 43:1686–1697
70. Langwinska-Wosko E, Skowronska M, Kmiec T, Czlonkowska A (2016) Retinal and optic nerve abnormalities in neurodegeneration associated with mutations in C19orf12 (MPAN). J Neurol Sci 370:237–240
71. Levi S, Finazzi D (2014) Neurodegeneration with brain iron accumulation: update on pathogenic mechanisms. Front Pharmacol 5:99
72. Levi S, Rovida E (2015) Neuroferritinopathy: from ferritin structure modification to pathogenetic mechanism. Neurobiol Dis 81:134–143
73. Maccarinelli F, Pagani A, Cozzi A, Codazzi F, Di Giacomo G, Capoccia S, Rapino S, Finazzi D, Politi LS, Cirulli F, Giorgio M, Cremona O, Grohovaz F, Levi S (2015) A novel neuroferritinopathy mouse model (FTL 498InsTC) shows progressive brain iron dysregulation, morphological signs of early neurodegeneration and motor coordination deficits. Neurobiol Dis 81:119–133
74. Maciel P, Cruz VT, Constante M, Iniesta I, Costa MC, Gallati S, Sousa N, Sequeiros J, Coutinho P, Santos MM (2005) Neuroferritinopathy: missense mutation in FTL causing early-onset bilateral pallidal involvement. Neurology 65:603–605
75. Malik I, Turk J, Mancuso DJ, Montier L, Wohltmann M, Wozniak DF, Schmidt RE, Gross RW, Kotzbauer PT (2008) Disrupted membrane homeostasis and accumulation of ubiquitinated proteins in a mouse model of infantile neuroaxonal dystrophy caused by PLA2G6 mutations. Am J Pathol 172:406–416
76. Mancuso M, Davidzon G, Kurlan RM, Tawil R, Bonilla E, Di Mauro S, Powers JM (2005) Hereditary ferritinopathy: a novel mutation, its cellular pathology, and pathogenetic insights. J Neuropathol Exp Neurol 64:280–294

77. Mari F, Berti B, Romano A, Baldacci J, Rizzi R, Grazia Alessandri M, Tessa A, Procopio E, Rubegni A, Lourenco CM, Simonati A, Guerrini R, Santorelli FM (2018) Clinical and neuroimaging features of autosomal recessive spastic paraplegia 35 (SPG35): case reports, new mutations, and brief literature review. Neurogenetics 19:123–130
78. McNeill A, Birchall D, Hayflick SJ, Gregory A, Schenk JF, Zimmerman EA, Shang H, Miyajima H, Chinnery PF (2008) T2* and FSE MRI distinguishes four subtypes of neurodegeneration with brain iron accumulation. Neurology 70:1614–1619
79. McNeill A, Chinnery PF (2011) Neurodegeneration with brain iron accumulation. Handb Clin Neurol 100:161–172
80. Miyajima H (2003) Aceruloplasminemia, an iron metabolic disorder. Neuropathology 23:345–350
81. Miyajima H (2015) Aceruloplasminemia. Neuropathology 35:83–90
82. Miyajima H, Nishimura Y, Mizoguchi K, Sakamoto M, Shimizu T, Honda N (1987) Familial apoceruloplasmin deficiency associated with blepharospasm and retinal degeneration. Neurology 37:761–767
83. Moutton S, Fergelot P, Trocello JM, Plante-Bordeneuve V, Houcinat N, Wenisch E, Larue V, Brugieres P, Clot F, Lacombe D, Arveiler B, Goizet C (2014) A novel FTL mutation responsible for neuroferritinopathy with asymmetric clinical features and brain anomalies. Parkinsonism Relat Disord 20:935–937
84. Nishida K, Garringer HJ, Futamura N, Funakawa I, Jinnai K, Vidal R, Takao M (2014) A novel ferritin light chain mutation in neuroferritinopathy with an atypical presentation. J Neurol Sci 342:173–177
85. Ogimoto M, Anzai K, Takenoshita H, Kogawa K, Akehi Y, Yoshida R, Nakano M, Yoshida K, Ono J (2011) Criteria for early identification of aceruloplasminemia. Intern Med 50:1415–1418
86. Ohta E, Nagasaka T, Shindo K, Toma S, Nagasaka K, Ohta K, Shiozawa Z (2008) Neuroferritinopathy in a Japanese family with a duplication in the ferritin light chain gene. Neurology 70:1493–1494
87. Orellana DI, Santambrogio P, Rubio A, Yekhlef L, Cancellieri C, Dusi S, Giannelli SG, Venco P, Mazzara PG, Cozzi A, Ferrari M, Garavaglia B, Taverna S, Tiranti V, Broccoli V, Levi S (2016) Coenzyme A corrects pathological defects in human neurons of PANK2-associated neurodegeneration. EMBO Mol Med 8:1197–1211
88. Panteghini C, Zorzi G, Venco P, Dusi S, Reale C, Brunetti D, Chiapparini L, Zibordi F, Siegel B, Garavaglia B, Simonati A, Bertini E, Nardocci N, Tiranti V (2012) C19orf12 and FA2H mutations are rare in Italian patients with neurodegeneration with brain iron accumulation. Semin Pediatr Neurol 19:75–81
89. Patel BN, Dunn RJ, Jeong SY, Zhu Q, Julien JP, David S (2002) Ceruloplasmin regulates iron levels in the CNS and prevents free radical injury. J Neurosci 22:6578–6586
90. Potter KA, Kern MJ, Fullbright G, Bielawski J, Scherer SS, Yum SW, Li JJ, Cheng H, Han X, Venkata JK, Khan PA, Rohrer B, Hama H (2011) Central nervous system dysfunction in a mouse model of FA2H deficiency. Glia 59:1009–1021
91. Powers JM (2006) p53-mediated apoptosis, neuroglobin overexpression, and globin deposits in a patient with hereditary ferritinopathy. J Neuropathol Exp Neurol 65:716–721
92. Proikas-Cezanne T, Waddell S, Gaugel A, Frickey T, Lupas A, Nordheim A (2004) WIPI-1alpha (WIPI49), a member of the novel 7-bladed WIPI protein family, is aberrantly expressed in human cancer and is linked to starvation-induced autophagy. Oncogene 23:9314–9325
93. Rajagopalan S, Rane A, Chinta SJ, Andersen JK (2016) Regulation of ATP13A2 via PHD2-HIF1alpha signaling is critical for cellular iron homeostasis: implications for Parkinson's disease. J Neurosci 36:1086–1095
94. Ramirez A, Heimbach A, Grundemann J, Stiller B, Hampshire D, Cid LP, Goebel I, Mubaidin AF, Wriekat AL, Roeper J, Al-Din A, Hillmer AM, Karsak M, Liss B, Woods CG, Behrens MI, Kubisch C (2006) Hereditary parkinsonism with dementia is caused by mutations in ATP13A2, encoding a lysosomal type 5 P-type ATPase. Nat Genet 38:1184–1191

95. Ramonet D, Podhajska A, Stafa K, Sonnay S, Trancikova A, Tsika E, Pletnikova O, Troncoso JC, Glauser L, Moore DJ (2012) PARK9-associated ATP13A2 localizes to intracellular acidic vesicles and regulates cation homeostasis and neuronal integrity. Hum Mol Genet 21:1725–1743
96. Roberti Mdo R, Borges Filho HM, Goncalves CH, Lima FL (2011) Aceruloplasminemia: a rare disease—diagnosis and treatment of two cases. Rev Bras Hematol Hemoter 33:389–392
97. Roth AD, Nunez MT (2016) Oligodendrocytes: functioning in a delicate balance between high metabolic requirements and oxidative damage. Adv Exp Med Biol 949:167–181
98. Roubertie A, Hieu N, Roux CJ, Leboucq N, Manes G, Charif M, Echenne B, Goizet C, Guissart C, Meyer P, Marelli C, Rivier F, Burglen L, Horvath R, Hamel CP, Lenaers G (2018) AP4 deficiency: a novel form of neurodegeneration with brain iron accumulation? Neurol Genet 4:e217
99. Salomao RP, Pedroso JL, Gama MT, Dutra LA, Maciel RH, Godeiro-Junior C, Chien HF, Teive HA, Cardoso F, Barsottini OG (2016) A diagnostic approach for neurodegeneration with brain iron accumulation: clinical features, genetics and brain imaging. Arq Neuropsiquiatr 74:587–596
100. Santambrogio P, Dusi S, Guaraldo M, Rotundo LI, Broccoli V, Garavaglia B, Tiranti V, Levi S (2015) Mitochondrial iron and energetic dysfunction distinguish fibroblasts and induced neurons from pantothenate kinase-associated neurodegeneration patients. Neurobiol Dis 81:144–153
101. Schneider SA, Bhatia KP (2012) Syndromes of neurodegeneration with brain iron accumulation. Semin Pediatr Neurol 19:57–66
102. Schultheis PJ, Fleming SM, Clippinger AK, Lewis J, Tsunemi T, Giasson B, Dickson DW, Mazzulli JR, Bardgett ME, Haik KL, Ekhator O, Chava AK, Howard J, Gannon M, Hoffman E, Chen Y, Prasad V, Linn SC, Tamargo RJ, Westbroek W, Sidransky E, Krainc D, Shull GE (2013) Atp13a2-deficient mice exhibit neuronal ceroid lipofuscinosis, limited alpha-synuclein accumulation and age-dependent sensorimotor deficits. Hum Mol Genet 22:2067–2082
103. Schulz K, Vulpe CD, Harris LZ, David S (2011) Iron efflux from oligodendrocytes is differentially regulated in gray and white matter. J Neurosci 31:13301–13311
104. Seibler P, Burbulla LF, Dulovic M, Zittel S, Heine J, Schmidt T, Rudolph F, Westenberger A, Rakovic A, Munchau A, Krainc D, Klein C (2018) Iron overload is accompanied by mitochondrial and lysosomal dysfunction in WDR45 mutant cells. Brain 141:3052–3064
105. Storti E, Cortese F, Di Fabio R, Fiorillo C, Pierallini A, Tessa A, Valleriani A, Pierelli F, Santorelli FM, Casali C (2013) De novo FTL mutation: a clinical, neuroimaging, and molecular study. Mov Disord 28:252–253
106. Texel SJ, Camandola S, Ladenheim B, Rothman SM, Mughal MR, Unger EL, Cadet JL, Mattson MP (2012) Ceruloplasmin deficiency results in an anxiety phenotype involving deficits in hippocampal iron, serotonin, and BDNF. J Neurochem 120:125–134
107. Venco P, Bonora M, Giorgi C, Papaleo E, Iuso A, Prokisch H, Pinton P, Tiranti V (2015) Mutations of C19orf12, coding for a transmembrane glycine zipper containing mitochondrial protein, cause mis-localization of the protein, inability to respond to oxidative stress and increased mitochondrial Ca(2)(+). Front Genet 6:185
108. Vidal R, Delisle MB, Rascol O, Ghetti B (2003) Hereditary ferritinopathy. J Neurol Sci 207:110–111
109. Vidal R, Ghetti B, Takao M, Brefel-Courbon C, Uro-Coste E, Glazier BS, Siani V, Benson MD, Calvas P, Miravalle L, Rascol O, Delisle MB (2004) Intracellular ferritin accumulation in neural and extraneural tissue characterizes a neurodegenerative disease associated with a mutation in the ferritin light polypeptide gene. J Neuropathol Exp Neurol 63:363–380
110. Vidal R, Miravalle L, Gao X, Barbeito AG, Baraibar MA, Hekmatyar SK, Widel M, Bansal N, Delisle MB, Ghetti B (2008) Expression of a mutant form of the ferritin light chain gene induces neurodegeneration and iron overload in transgenic mice. J Neurosci 28:60–67
111. Wanders RJ, Jansen GA, Lloyd MD (2003) Phytanic acid alpha-oxidation, new insights into an old problem: a review. Biochim Biophys Acta 1631:119–135

112. Yamamoto K, Yoshida K, Miyagoe Y, Ishikawa A, Hanaoka K, Nomoto S, Kaneko K, Ikeda S, Takeda S (2002) Quantitative evaluation of expression of iron-metabolism genes in ceruloplasmin-deficient mice. Biochim Biophys Acta 1588:195–202
113. Yang Y, Wu Z, Kuo YM, Zhou B (2005) Dietary rescue of fumble—a Drosophila model for pantothenate-kinase-associated neurodegeneration. J Inherit Metab Dis 28:1055–1064
114. Yoshida K, Furihata K, Takeda S, Nakamura A, Yamamoto K, Morita H, Hiyamuta S, Ikeda S, Shimizu N, Yanagisawa N (1995) A mutation in the ceruloplasmin gene is associated with systemic hemosiderosis in humans. Nat Genet 9:267–272
115. Zanardi A, Conti A, Cremonesi M, D'Adamo P, Gilberti E, Apostoli P, Cannistraci CV, Piperno A, David S, Alessio M (2018) Ceruloplasmin replacement therapy ameliorates neurological symptoms in a preclinical model of aceruloplasminemia. EMBO Mol Med 10:91–106
116. Zhao Z, Zhang X, Zhao C, Choi J, Shi J, Song K, Turk J, Ma ZA (2010) Protection of pancreatic beta-cells by group VIA phospholipase A(2)-mediated repair of mitochondrial membrane peroxidation. Endocrinology 151:3038–3048
117. Zinoviev A, Goyal A, Jindal S, LaCava J, Komar AA, Rodnina MV, Hellen CUT, Pestova TV (2018) Functions of unconventional mammalian translational GTPases GTPBP1 and GTPBP2. Genes Dev 32:1226–1241
118. Zizioli D, Tiso N, Guglielmi A, Saraceno C, Busolin G, Giuliani R, Khatri D, Monti E, Borsani G, Argenton F, Finazzi D (2016) Knock-down of pantothenate kinase 2 severely affects the development of the nervous and vascular system in zebrafish, providing new insights into PKAN disease. Neurobiol Dis 85:35–48
119. Zorzi G, Zibordi F, Chiapparini L, Bertini E, Russo L, Piga A, Longo F, Garavaglia B, Aquino D, Savoiardo M, Solari A, Nardocci N (2011) Iron-related MRI images in patients with pantothenate kinase-associated neurodegeneration (PKAN) treated with deferiprone: results of a phase II pilot trial. Mov Disord 26:1756–1759

Chapter 10
Diagnostics and Treatments of Iron-Related CNS Diseases

Huan Xiong, Qing-zhang Tuo, Yu-jie Guo and Peng Lei

Abstract Iron has been proposed to be responsible for neuronal loss in several diseases of the central nervous system, including Alzheimer's disease (AD), Parkinson's disease (PD), stroke, Friedreich's ataxia (FRDA), multiple sclerosis (MS), amyotrophic lateral sclerosis (ALS). In many diseases, abnormal accumulation of brain iron in disease-affected area has been observed, without clear knowledge of the contribution of iron overload to pathogenesis. Recent evidences implicate that key proteins involved in the disease pathogenesis may also participate in cellular iron metabolism, suggesting that the imbalance of brain iron homeostasis is associated with the diseases. Considering the complicated regulation of iron homeostasis within the brain, a thorough understanding of the molecular events leading to this phenotype is still to be investigated. However, current understanding has already provided the basis for the diagnosis and treatment of iron-related CNS diseases, which will be reviewed here.

Keywords Iron · Alzheimer's disease · Iron homeostasis · Parkinson's disease · Stroke

10.1 Introduction

In the central nervous system (CNS), iron is involved in multiple crucial biological processes including oxygen transportation, DNA synthesis, mitochondrial respiration, myelin synthesis, and the synthesis and metabolism of neurotransmitters [1]. It plays an important role in electron transfer and is cofactors for varies enzymes [2]. Iron homoeostasis is needed to maintain normal physiological brain function,

H. Xiong · Q. Tuo · P. Lei (✉)
Department of Neurology and State Key Laboratory of Biotherapy, West China Hospital, Sichuan University, Sichuan 610041, China
e-mail: peng.lei@scu.edu.cn

Y. Guo · P. Lei
West China School of Basic Medical Sciences and Forensic Medicine, Sichuan University, Sichuan 610041, China

Y.-Z. Chang (ed.), *Brain Iron Metabolism and CNS Diseases*, Advances in Experimental Medicine and Biology 1173,
https://doi.org/10.1007/978-981-13-9589-5_10

whereas abnormal iron homoeostasis can induce cellular damage through reactive oxygen species (ROS), notably the hydroxyl radical [3]. ROS can cause the oxidation and modification of lipids, proteins, carbohydrates, and DNA [4, 5]. Peroxidation of polyunsaturated fatty acids in membrane lipids by ROS can generate highly reactive aldehydes, such as 4-hydroxynonenal, which irreversibly modify proteins by carbonylation [6–8]. ROS can induce the release of iron from mitochondrial iron-sulfur cluster proteins of the respiratory chain and other iron storage proteins, which will lead to further ROS production via Fenton's reaction [9]. Iron-related CNS diseases that result from iron toxicity can lead to apoptosis [10] and ferroptosis [11], an iron-specific form of non-apoptotic cell death [12].

In various iron-related CNS diseases, changes in iron homoeostasis result in altered cellular iron distribution and accumulation [13–21]. Magnetic resonance imaging (MRI) can often identify these changes, thus providing a potential diagnostic biomarker of iron-related CNS diseases. An important avenue to reduce iron accumulation is the use of iron chelators that are able to cross the blood–brain barrier, penetrate cells, and reduce excessive iron accumulation, thereby affording neuroprotection. These will provide new ideas for the diagnostics and treatment of iron-related CNS diseases.

10.2 Potential Diagnostic Value of Iron for CNS Diseases

10.2.1 Alzheimer's Disease (AD)

Making a diagnosis of AD on purely clinical grounds is challenging [22], not only in the prodromal stage in which patients only have subtle cognitive symptoms but also in the dementia phase. With the recognition that the pathological changes occur years prior to symptoms, and the advent of biomarkers of β-amyloid (Aβ) and tau pathology and MRI measures of atrophy, diagnostic criteria have evolved to allow for the diagnosis to be made both earlier and with increased molecular specificity. The most recent diagnostic criteria from both the National Institute of Aging (NIA) and the International Working Group (IWG-2) now incorporate one or more preclinical AD phases, where biomarker evidence of AD pathology exists in the absence of symptoms [23–25].

The mainstay of the diagnosis of AD remains the clinical assessment, and in particular the clinical interview with the patient and an informant, and a cognitive and focused physical examination. Neuropsychology allows for the quantification of both the pattern and severity of cognitive deficits against age-related norms.

Latest MRI showed that iron levels in the bilateral hippocampus (HP), parietal cortex (PC), frontal white matter, putamen (PU), caudate nucleus (CN), and dentate nucleus (DN) subregions of patients with AD were significantly higher than the healthy controls. Moreover, these brain iron levels, especially those in the PC at the early stages of AD, were positively correlated with the severity of patients' cognitive impairment [26]. These data indicated that iron in specific brain regions may be used as a biomarker to evaluate the progression of AD. Australian researchers have applied quantitative susceptibility mapping (QSM), a relatively new MRI method sensitive to tissue iron, to assess the relationship between iron, Aβ load, and cognitive decline. They found that QSM level is associated with Aβ deposition, which is consistent with brain iron increasing during the natural history of AD [27]. This study showed that higher levels of brain iron can also predict clinical deterioration.

Ferritin is the major iron storage protein of the body; by using cerebrospinal fluid (CSF) levels of ferritin as an index, researchers explored whether brain iron status impacts the longitudinal outcomes in the AD Neuroimaging Initiative (ADNI) cohort. Results showed that CSF ferritin levels were negatively associated with cognitive performance over 7 year's period and predicted the conversion from mild cognitive impairment (MCI) to AD [28]. These findings reveal that ferritin levels in the CSF may predict AD outcomes.

10.2.2 Parkinson's Disease (PD)

Parkinson's disease is a common neurodegenerative disease that mainly affects the motor system appears as tremor, rigidity, and bradykinesia, and especially involved people over 60 years old [29]. The cause of PD is unknown, and there is no effective means to cure PD. The primary pathology of PD is dopaminergic neuron loss in substantia nigra (SN). The standard of diagnosis PD mainly included assessment using Unified PD Rating Scale (UPDRS) motor score based on the motor symptoms, but a single-photon emission computerized tomography (SPECT) scan for dopamine transporter may also be helpful. In addition, iron may play an important role in PD.

An epidemic research proceeded by Powers et al. showed that dietary iron intake exhibited a positive dose–response gradient with PD risk in men, through investigated newly diagnosed Parkinson's disease case ($n = 420$) and normal control ($n = 560$) in Washington State with Willett Food Frequency Questionnaire (FFQ) [30]. Whereas a Mendelian randomization research about 20000 PD patients and 80000 normal controls had shown that increased of serum iron level would reduce the risk of PD [31], while the CSF level of iron was not changed in PD patients in a research about 37 PD patients [32], the mechanism underlying remains unclear.

Transcranial sonography (TCS) and later T_2^*-weighted MRI serve as general technologies clinically to detect iron in SN of PD, for the laden iron will increase echogenicity in these technologies [33, 34], the positive predictive value is 85.7% [35]. In a 3-year clinical study on PD patients and normal controls, the iron content measured by T_2^*-weighted MRI is increased in SN and caudal putamen in PD group while not changed in control [35]. In addition, Zhang et al. found that iron concentration in SN of PD patients was a positive correlation with the UPDRS of these PD patients by using high-resolution susceptibility-weighted magnetic resolution imaging [36].

Some studies in postmortem PD brains also get the similar conclusion that iron is overloaded in SN in PD. For instance, detection the concentration of metal ion in SN with Electrothermal Atomic Absorption Spectrometry (ETAAS), discovered that the labile iron concentration but not copper was significantly higher in PD patients than in healthy control [37]. Lei et al. also found that iron elevation in PD patients and demonstrated that it may be associated with tau deficiency [15].

Diseases that affect iron status also have influences on PD. Mutation of gene encoding ferritin light polypeptide would cause iron deposition and lead to adult-onset basal ganglia disease such as parkinsonism symptom [38, 39]. Ceruloplasmin (Cp) is a copper oxidase and plays a role in iron export from cell [40]. About 176 PD patients from UK Brain Bank with increased SN iron level were examined for mutations of Cp gene and found that five missense variations about Cp were related to PD [41]. Mutation of pantothenate kinase 2 (PANK2) will lead to iron accumulation in basal ganglia [42], while two siblings with this PANK2 mutation come from consanguineous parents have developed parkinsonism [43].

10.2.3 Stroke

Stroke is a leading cause of mortality and permanent disability [44]. In general, the stroke is classicized as ischemic and hemorrhagic stroke. Ischemic stroke is the most common type, accounting for about 80% of the total stroke events [45]. Clinical studies indicated that there was iron-mediated neurotoxicity in patients with ischemic stroke. The increased iron deposition, measured by MRI, was found in the basal ganglia, thalami, and white matter in children with severe ischemic-anoxic insult and subsequent resuscitation [17]. In addition, the serum transferrin, a main protein regulating iron homeostasis, and ceruloplasmin were correlated with the ischemic lesion volume and hydroperoxides, respectively [46]. High level of serum ferritin has been found to be associated with poor outcome in patients with ischemic stroke [47]. In further research, plasma ferritin concentrations >275 ng/ml with 24 h from the onset of ischemic stroke increased 33-fold the risk of subsequent stroke progression [48]. These results indicated that the systemic iron level may be associated with the clinical states of ischemic stroke.

In a study of hemorrhagic stroke, there was a significant correlation between the serum ferritin level and the relative perihematoma edema volume at 3–4 days after hemorrhagic stroke onset [49]. The serum ferritin level was upregulated after hemorrhagic stroke, and the higher levels of serum ferritin, the poorer outcome in patients [50, 51]. Another clinical study showed that the hematoma iron content, measured by MRI, correlated with the relative perihematoma edema volume [52]. Therefore, iron-related signals and tests may help the diagnosis of stroke.

10.2.4 Friedreich's Ataxia (FRDA)

Most of FRDA patients caused by trinucleotide amplification of GAA in frataxin (FXN) gene [53]. The FXN protein, an iron chaperone in iron–sulfur cluster, is related to iron accumulation in mitochondria [54]. Although a conclusive diagnosis of FRDA can only be made via genetic testing, other helpful tests, such as the iron concentrations and the expression of iron-related proteins, may be performed to be a potential biomaker of FRDA. Clinical study showed the iron accumulation in the dentate nucleus and the dorsal root ganglia of FRDA patients [18, 55, 56]. And the expression of iron-related proteins changed in individuals with FRDA [57]. The mitochondrial iron–sulfur-cluster-containing enzymes were found to be deficient by biochemical studies of heart biopsies [58]. Compared with control neurons, FRDA neurons exhibited higher labile iron pool, reactive oxygen species, and lower reduced GSH levels, as well as enhanced sensitivity to oxidants [59]. In a yeast model of FRDA, GSH peroxidase activity increased with the reduction of total GSH levels [60].

10.2.5 Restless Legs Syndrome (RLS)

RLS is a chronic neurological disorder with unique clinical symptoms. The diagnosis of RLS used to be based on the subjective description of symptoms, lacking of available biomarkers or objective tests [61]. Since RLS can be exacerbated by systemic iron deficiency, all newly diagnosed patients as well as those with worsening symptoms should have iron-level tests [61]. Clinical study showed that the SN iron index, determined by MRI, significantly decreased in the early-onset RLS patients. And there was a significant negative correlation between SN iron index and the Johns Hopkins RLS severity scale [62]. In addition, iron deficiency was found in red nucleus, pallidum, thalamus, and putamen of RLS patients [63, 64]. Furthermore, compared to healthy control, the CSF ferritin and transferrin levels from RLS patients significantly decreased and increased, respectively [65].

10.2.6 Multiple Sclerosis (MS)

The diagnosis of MS needs the integration of clinical, imaging, and laboratory evidences. Besides, the magnetic field correlations value, indicating iron deposition, was increased in the putamen, thalamus, and globus pallidus in MS patients compared with the healthy control [66]. And MS lesions had 47 μg of iron per gram of tissue more than the surrounding tissue on average, quantified by susceptibility-weighted imaging [67]. In chronic MS, iron levels reduced in normal-appearing white matter and correlated with disease duration. The ferritin and frataxin levels were upregulated and downregulated in initial lesions, respectively [68]. Previous studies indicated that the level of iron deposition might be related to disease duration and severity [69, 70].

10.2.7 Amyotrophic Lateral Sclerosis (ALS)

ALS, also known as 'Lou Gehrig's disease,' is a neurodegenerative disorder [71]. The elevation of iron levels has been found in the motor cortex, the spinal neurons, and the CSF of patients with ALS, as well as in the spinal cord in mouse models of ALS [19, 72–74]. The expression of several iron regulators changed in ALS, such as mitochondrial ferritin, TfR1, DMT1, and FPN1 [75]. Moreover, the CSF ferric reducing ability decreased [76] and serum ferritin levels increased in ALS patients [77]. Interestingly, transferrin has been observed in the Bunina bodies which are the typical pathological symptoms of ALS. Besides cystatin C, transferrin is the only protein localized in the Bunina bodies [78]. Furthermore, mutation of HFE gene, associated with the iron overload disease and hemochromatosis, increased in ALS patients [79].

10.3 Treatment of Iron-Related CNS Diseases

10.3.1 Alzheimer's Disease

Aging is the single highest risk factor for AD with an incidence of 50% in people over the age of 85 [80]. Over more than thirty years of research, no disease-modifying treatment has been approved by either the US Food and Drug Administration (FDA) or the European Medicines Agency (EMA) for AD treatment.

Given the pathological iron accumulation in AD, iron removal from specific brain region without causing a systematic iron deficiency should be taken into consideration [81, 82]. McLachlan and colleagues [83] conducted a two-year, single-blinded study to investigate whether the progression of dementia could be slowed by an iron chelator, desferrioxamine (DFO), and showed a substantial reduction in the rate of deterioration of daily living skills in 48 patients with Alzheimer's disease who were given DFO (125 mg intramuscularly twice daily, 5 days per week, for 24 months) when compared with Alzheimer's disease given placebo. Despite the promising results, no other clinical trials using DFO for AD have been investigated so far. And yet, iron chelation using intranasal deferoxamine has been investigated in mouse models of AD, was shown to reverse iron-induced memory deficits, and inhibits amyloidogenic APP processing and Aβ aggregation [84–86].

5-Chloro-7-iodoquinolin-8-ol (clioquinol) is a moderate chelator of iron, copper, and zinc, and for this reason, it was explored as a drug candidate for AD [87]. Indeed, clioquinol has shown efficacy in patients with AD and models of diseases. Clioquinol inhibits Aβ oligomer formation and protects against cell loss in an Aβ-injection model [88–90]. Nine weeks of oral clioquinol treatment lowered plaque burden and improved cognitive performance in AD mice [91, 92]. In a phase 2 clinical trial of 32 patients, clioquinol prevented cognitive deterioration (Alzheimer's Disease Assessment Scale-Cognitive) and lowered plasma Aβ42 levels over 36 weeks [93]. While these data were promising, complications with the large-scale manufacturing of the compound made further development of this drug unviable.

PBT2 (5,7-dichloro-2-[(dimethylamino)-methyl]-8-hydroxyquinoline), the second generation of clioquinol derivative, has been shown to promote amyloid plaque degradation and relocate excess released metals to other depleted neuronal compartments [94–96]. Given orally to two types of amyloid-bearing transgenic mouse models of AD (APP/PS1 transgenic mice and Tg2576 transgenic mice), PBT2 outperformed clioquinol by markedly decreasing soluble interstitial brain Aβ within hours and improving cognitive performance to exceed that of normal littermate controls within days [94]. In a phase 2a clinical trial of 78 patients with Alzheimer's disease, PBT2 was given at a dose of 250 mg/day or 50 mg/day, and patients receiving highest dose showed a substantial reduction in CSF of Aβ concentration, and some cognitive improvement (executive function) was noted [97, 98].

A randomized, multi-center, double-blinded, placebo-controlled Phase 2a trial using deferiprone (DFP) for AD is currently ongoing in Melbourne (Deferiprone to Delay Dementia, the 3D study), and the results are due in 2021.

10.3.2 Parkinson's Disease

The primary treatment strategies for Parkinson's disease include medications such as carbidopa-levodopa, dopamine agonists, MAO-B inhibitors, as well as anticholinergics, and deep brain stimulation surgery if necessary. Given the substantial evidences linking iron accumulation with PD, a few treatments targeting iron were trialed.

The iron chelator deferiprone had been trialed for PD. A patient with symptoms of dysarthria, orofacial dystonia, moderate parkinsonism, iron accumulation in the substantia nigra, red nuclei, internal globi pallidi, and dentate nuclei detected by T2*-weighted brain MRI had developed worsen during 5 years even was given medications of levodopa, baclofen, oxitriptan, and buspirone. She was given 30 mg/kg deferiprone per day for 32 months and found that her symptoms were improved 30% after 6 months treatment, and the UPDRS decreased 30% after 1 year, while iron accumulation in bilateral nuclei, substantia nigra were decreased, but not changed in the red nuclei after 32 months [99]. PD is associated with iron deposition in SN and iron deposition will increase oxidative damage. Grolez et al. carried out a pilot clinical trial about treating PD patients with deferiprone or placebo for 6–12 months and found that the iron level in SN and the UPDSR was significantly decreased; in addition, if the patients have gene mutation about iron metabolism, the effect of reducing iron was stronger [100]. Devos et al. detected cytoprotection of deferiprone treated cell models of PD or MPTP-inject mouse model of PD, and found a high level cytoprotection of deferiprone. Whereafter they did a randomized clinical trial on early-stage PD patients whom were treated with deferiprone for 12 months, revealed that deferiprone definite lower the iron deposition and UPDRS, and the effect would come earlier when the treat begin sooner [101].

Except deferiprone, the efficacy of other iron chelators or iron modulation therapy for PD has been investigated in animal models of PD. Tau knockout mice which develop iron deposition in SN and parkinsonism motor disorder would have reduced iron content in the brain and reversed motor deficit after 5-month clioquinol treatment [102]. In PD, Cp activity decrease may contribute to the iron accumulation, and peripheral infusion of Cp weakens iron deposition in SN of PD mouse model [103]. VAR10303 (VAR) was a brain-permeable iron chelator attached with lipoperoxidation inhibitory potency, and animal models of PD include 6-hydroxydopamine stereotactically injected rats and 1-methyl-4-phenyl-1,2,3,6-tetrahydropyridine intraperitoneally injected mice were treated with VAR and revealed that VAR can increase tyrosine-hydroxylase level and improve synaptic plasticity and some other neuroprotective and neurorestorative effects [104]. Multifunctional iron chelators similar to VAR, such as VK-28 and M30, have also been designed and tested in PD models in vitro or in vivo [105–107].

Interestingly, Kaur et al. treated MPTP-inject mouse model of PD with transgenic expression of ferritin, detected reduced oxidative stress of dopaminergic neurons in SN, and found genetic and pharmacological means both efficient to protect dopaminergic neurons [108].

10.3.3 Stroke

Injection of tissue plasminogen activator (tPA, also called alteplase) is the only evidence-based treatment option for ischemic stroke [109]. However, limited by the narrow therapeutic window (less than 4.5 h after the onset of ischemic stroke) and other safety concerns such as potential bleeding in the brain, less than 5% of patients have received this treatment [110]. Therefore, other effective therapies are needed. Deferoxamine mesylate has exhibited neuroprotective effects by suppression of iron-induced hydroxyl radical formation. In a clinical trial, the serum levels of peroxides had decreased and the levels of total radical-trapping antioxidant capacity had increased after 3-day treatment with deferoxamine mesylate in stroke patients [111]. Tirilazad mesylate, a nonglucocorticoid, 21-aminosteroid, and iron-dependent peroxidation inhibitor, has been investigated. In a randomized, double-blinded, placebo-controlled trial, patients with acute stroke were treated with the tirilazad mesylate beginning at a median of 4.3 h after stroke. While the overall functional outcome had not been improved measured by the Glasgow Outcome Scale and the Barthel Index 3 months later [112], a higher dose showed 14% absolute reduction in mortality [113]. In a rat model of ischemic stroke, intranasally administered deferoxamine improved the outcome [114], and inhibitors of ferroptosis protected from ischemic injury in MCAO mouse model [115].

As to treat hemorrhagic stroke, five randomized, placebo-controlled trials were performed, and the tirilazad had not shown effective on clinical outcome, but decreased symptomatic vasospasm in patients with subarachnoid hemorrhage [116].

10.3.4 Friedreich's Ataxia (FRDA)

Since excess mitochondrial iron may worse the symptom of FRDA, iron chelators were considered to treat FRDA. Deferoxamine cannot cross the blood–brain barrier [117] and therefore might induce extracellular iron deficiency before the reduction of mitochondrial iron [118]. Another orally iron chelator deferiprone can cross the blood–brain barrier with low affinity to iron [117]. Clinical trials indicated that deferiprone at low concentrations had beneficial effects, but worse the symptoms of FRDA at high doses [55, 119]. In another clinical trial, deferiprone and idebenone, latter of which was an analog of coenzyme Q_{10}, were combined to treat FRDA. After 11 months of therapy, iron deposits in dentate nucleus, determined by MRI, as well as heart hypertrophy parameters, were improved significantly. However, posture scores and gait got worse [120].

10.3.5 Restless Legs Syndrome (RLS)

Clinical studies suggested that in certain cases correcting the iron deficiency by iron supplementation alone can relieve RLS without any further intervention [61]. In an open-label study, a single 1000 mg intravenous (IV) infusion of iron dextran to treat RLS was evaluated. The results indicated that IV iron dextran significantly improved the RLS symptom severity, duration, and total sleep time two weeks later. In addition, brain iron concentration as determined by MRI was increased in the SN and the prefrontal cortex [121]. In a randomized, parallel-group double-blinded study, IV iron sucrose treatment significantly increased the CSF ferritin level and reduced the RLS symptom severity, but did not change MRI iron index or periodic leg movements of sleep [122].

10.3.6 Multiple Sclerosis (MS)

Although iron chelation therapy works in animal model of MS [123], clinical studies have failed to show its efficacy. In one clinical study, desferrioxamine was given to 9 patients up to 8 courses over 2 years. During the trial, 1 patient was improved, 3 patients unchanged, and 5 patients got worse, although the drug was well tolerated [124, 125]. A preliminary data indicated that the symptoms of 2–4 patients with progressive MS could be improved by the combination of iron depletion and erythropoietin [126]. Therefore, the effectiveness of iron chelators for MS needs further investigation.

10.3.7 Amyotrophic Lateral Sclerosis (ALS)

Recently, iron chelation for ALS was assessed in a pilot clinical trial. The results showed that the body mass index and the ALS Functional Rating Scale significantly decreased for the first 3 months of deferiprone therapy. The iron levels in the cervical spinal cord, motor cortex, and medulla oblongata reduced after deferiprone treatment. These data encourage further attempt to test iron chelation for ALS.

10.4 Conclusion

Evidences have shown that iron deposition in brain was closely related with the pathogenesis of multiple CNS diseases, and treatments to reduce iron in brain have been investigated in clinical and animal studies. With positive outcomes, there are still challenges to use iron chelation as therapy for brain disorders. Therefore, future investigations are required to develop iron-trapping agents that are highly selective, low affinity, BBB permeable, low in toxicity. With the right agent, the right time window, and the right patients to intervene, iron will be a tractable target for CNS diseases.

References

1. Crichton RR, Boelaert JR (2001) Inorganic biochemistry of iron metabolism: from molecular mechanisms to clinical consequences. 2nd edn. Wiley
2. Rouault TA (2013) Iron metabolism in the CNS: implications for neurodegenerative diseases. Nat Rev Neurosci 14:551–564. https://doi.org/10.1038/nrn3453
3. Crichton RR, Ward RJ (2008) Metal-based neurodegeneration: from molecular mechanisms to therapeutic strategies. 2nd edn
4. Hentze MW, Muckenthaler MU, Galy B, Camaschella C (2010) Two to tango: regulation of mammalian iron metabolism. Cell 142:24–38. https://doi.org/10.1016/j.cell.2010.06.028
5. Wang J, Pantopoulos K (2011) Regulation of cellular iron metabolism. Biochem J 434:365–381. https://doi.org/10.1042/BJ20101825
6. Catala A (2009) Lipid peroxidation of membrane phospholipids generates hydroxy-alkenals and oxidized phospholipids active in physiological and/or pathological conditions. Chem Phys Lipids 157:1–11. https://doi.org/10.1016/j.chemphyslip.2008.09.004
7. Dalle-Donne I, Giustarini D, Colombo R, Rossi R, Milzani A (2003) Protein carbonylation in human diseases. Trends Mol Med 9:169–176
8. Perluigi M, Coccia R, Butterfield DA (2012) 4-Hydroxy-2-nonenal, a reactive product of lipid peroxidation, and neurodegenerative diseases: a toxic combination illuminated by redox

proteomics studies. Antioxid Redox Signal 17:1590–1609. https://doi.org/10.1089/ars.2011.4406
9. Raha S, Robinson BH (2000) Mitochondria, oxygen free radicals, disease and ageing. Trends Biochem Sci 25:502–508
10. Ott M, Gogvadze V, Orrenius S, Zhivotovsky B (2007) Mitochondria, oxidative stress and cell death. Apoptosis 12:913–922. https://doi.org/10.1007/s10495-007-0756-2
11. Stockwell BR et al (2017) Ferroptosis: a regulated cell death nexus linking metabolism, redox biology, and disease. Cell 171:273–285. https://doi.org/10.1016/j.cell.2017.09.021
12. Dixon SJ et al (2012) Ferroptosis: an iron-dependent form of nonapoptotic cell death. Cell 149:1060–1072. https://doi.org/10.1016/j.cell.2012.03.042
13. Zecca L, Youdim MB, Riederer P, Connor JR, Crichton RR (2004) Iron, brain ageing and neurodegenerative disorders. Nat Rev Neurosci 5:863–873. https://doi.org/10.1038/nrn1537 [pii]
14. Ward RJ, Zucca FA, Duyn JH, Crichton RR, Zecca L (2014) The role of iron in brain ageing and neurodegenerative disorders. Lancet Neurol 13:1045–1060. https://doi.org/10.1016/S1474-4422(14)70117-6
15. Lei P et al (2012) Tau deficiency induces parkinsonism with dementia by impairing APP-mediated iron export. Nat Med 18:291–295. https://doi.org/10.1038/nm.2613
16. Duce JA et al (2010) Iron-export ferroxidase activity of β-amyloid precursor protein is inhibited by zinc in Alzheimer's disease. Cell 142:857–867. https://doi.org/10.1016/j.cell.2010.08.014
17. Dietrich RB, Bradley WG Jr (1988) Iron accumulation in the basal ganglia following severe ischemic-anoxic insults in children. Radiology 168:203–206. https://doi.org/10.1148/radiology.168.1.3380958
18. Waldvogel D, van Gelderen P, Hallett M (1999) Increased iron in the dentate nucleus of patients with Friedrich's ataxia. Ann Neurol 46:123–125
19. Kwan JY et al (2012) Iron accumulation in deep cortical layers accounts for MRI signal abnormalities in ALS: correlating 7 tesla MRI and pathology. PLoS ONE 7:e35241. https://doi.org/10.1371/journal.pone.0035241
20. Ropele S et al (2011) MRI assessment of iron deposition in multiple sclerosis. J Magn Reson Imaging 34:13–21. https://doi.org/10.1002/jmri.22590
21. Allen RP, Barker PB, Wehrl FW, Song HK, Earley CJ (2001) MRI measurement of brain iron in patients with restless legs syndrome. Neurology 56:263–265
22. Knopman DS et al (2001) Practice parameter: diagnosis of dementia (an evidence-based review). Report of the Quality Standards Subcommittee of the American Academy of Neurology. Neurology 56:1143–1153
23. Sperling RA et al (2011) Toward defining the preclinical stages of Alzheimer's disease: recommendations from the National Institute on Aging-Alzheimer's Association workgroups on diagnostic guidelines for Alzheimer's disease. Alzheimers Dement 7:280–292. https://doi.org/10.1016/j.jalz.2011.03.003
24. Dubois B et al (2014) Advancing research diagnostic criteria for Alzheimer's disease: the IWG-2 criteria. Lancet Neurol 13:614–629. https://doi.org/10.1016/S1474-4422(14)70090-0
25. Dubois B et al (2016) Preclinical Alzheimer's disease: definition, natural history, and diagnostic criteria. Alzheimers Dement 12:292–323. https://doi.org/10.1016/j.jalz.2016.02.002
26. Zhu WZ et al (2009) Quantitative MR phase-corrected imaging to investigate increased brain iron deposition of patients with Alzheimer disease. Radiology 253:497–504. https://doi.org/10.1148/radiol.2532082324
27. Ayton S et al (2017) Cerebral quantitative susceptibility mapping predicts amyloid-beta-related cognitive decline. Brain 140:2112–2119. https://doi.org/10.1093/brain/awx137
28. Ayton S, Faux NG, Bush AI, Alzheimer's Disease Neuroimaging I (2015) Ferritin levels in the cerebrospinal fluid predict Alzheimer's disease outcomes and are regulated by APOE. Nat Commun 6:6760. https://doi.org/10.1038/ncomms7760
29. de Lau LM, Breteler MM (2006) Epidemiology of Parkinson's disease. Lancet Neurol 5:525–535. https://doi.org/10.1016/S1474-4422(06)70471-9

30. Powers KM et al (2009) Dietary fats, cholesterol and iron as risk factors for Parkinson's disease. Parkinsonism Relat Disord 15:47–52. https://doi.org/10.1016/j.parkreldis.2008.03.002
31. Pichler I et al (2013) Serum iron levels and the risk of Parkinson disease: a Mendelian randomization study. PLoS Med 10:e1001462. https://doi.org/10.1371/journal.pmed.1001462
32. Jimenez-Jimenez FJ et al (1998) Cerebrospinal fluid levels of transition metals in patients with Parkinson's disease. J Neural Transm (Vienna) 105:497–505. https://doi.org/10.1007/s007020050073
33. Zecca L et al (2005) In vivo detection of iron and neuromelanin by transcranial sonography: a new approach for early detection of substantia nigra damage. Mov Disord 20:1278–1285. https://doi.org/10.1002/mds.20550
34. Becker G, Seufert J, Bogdahn U, Reichmann H, Reiners K (1995) Degeneration of substantia nigra in chronic Parkinson's disease visualized by transcranial color-coded real-time sonography. Neurology 45:182–184
35. Prestel J, Schweitzer KJ, Hofer A, Gasser T, Berg D (2006) Predictive value of transcranial sonography in the diagnosis of Parkinson's disease. Mov Disord 21:1763–1765. https://doi.org/10.1002/mds.21054
36. Zhang J et al (2010) Characterizing iron deposition in Parkinson's disease using susceptibility-weighted imaging: an in vivo MR study. Brain Res 1330:124–130. https://doi.org/10.1016/j.brainres.2010.03.036
37. Wypijewska A et al (2010) Iron and reactive oxygen species activity in parkinsonian substantia nigra. Parkinsonism Relat Disord 16:329–333. https://doi.org/10.1016/j.parkreldis.2010.02.007
38. Curtis AR et al (2001) Mutation in the gene encoding ferritin light polypeptide causes dominant adult-onset basal ganglia disease. Nat Genet 28:350–354. https://doi.org/10.1038/ng571
39. Maciel P et al (2005) Neuroferritinopathy: missense mutation in FTL causing early-onset bilateral pallidal involvement. Neurology 65:603–605. https://doi.org/10.1212/01.wnl.0000178224.81169.c2
40. Osaki S, Johnson DA, Frieden E (1966) The possible significance of the ferrous oxidase activity of ceruloplasmin in normal human serum. J Biol Chem 241:2746–2751
41. McNeill A, Pandolfo M, Kuhn J, Shang H, Miyajima H (2008) The neurological presentation of ceruloplasmin gene mutations. Eur Neurol 60:200–205. https://doi.org/10.1159/000148691
42. Swaiman KF (1991) Hallervorden-Spatz syndrome and brain iron metabolism. Arch Neurol 48:1285–1293
43. Alberca R, Rafel E, Chinchon I, Vadillo J, Navarro A (1987) Late onset parkinsonian syndrome in Hallervorden-Spatz disease. J Neurol Neurosurg Psychiatry 50:1665–1668
44. Krishnamurthi RV et al (2013) Global and regional burden of first-ever ischaemic and haemorrhagic stroke during 1990–2010: findings from the global burden of disease study 2010. Lancet Global Health 1:E259–E281
45. Langhorne P, Bernhardt J, Kwakkel G (2011) Stroke rehabilitation. Lancet 377:1693–1702. https://doi.org/10.1016/S0140-6736(11)60325-5
46. Altamura C et al (2009) Ceruloplasmin/transferrin system is related to clinical status in acute stroke. Stroke: J Cereb Circ 40:1282–1288. https://doi.org/10.1161/STROKEAHA.108.536714
47. Davalos A et al (1994) Iron-related damage in acute ischemic stroke. Stroke: J Cereb Circ 25:1543–1546
48. Davalos A et al (2000) Body iron stores and early neurologic deterioration in acute cerebral infarction. Neurology 54:1568–1574
49. Mehdiratta M, Kumar S, Hackney D, Schlaug G, Selim M (2008) Association between serum ferritin level and perihematoma edema volume in patients with spontaneous intracerebral hemorrhage. Stroke: J Cereb Circ 39:1165–1170. https://doi.org/10.1161/STROKEAHA.107.501213
50. Perez de la Ossa, N et al (2010) Iron-related brain damage in patients with intracerebral hemorrhage. Stroke: J Cereb Circ 41:810–813. https://doi.org/10.1161/strokeaha.109.570168

51. Xiong XY, Wang J, Qian ZM, Yang QW (2014) Iron and intracerebral hemorrhage: from mechanism to translation. Transl Stroke Res 5:429–441. https://doi.org/10.1007/s12975-013-0317-7
52. Lou M, Lieb K, Selim M (2009) The relationship between hematoma iron content and perihematoma edema: an MRI study. CerebVascular Dis 27:266–271. https://doi.org/10.1159/000199464
53. Wu JR, Tuo QZ, Lei P (2018) Ferroptosis, a recent defined form of critical cell death in neurological disorders. J Mol Neurosci: MN. https://doi.org/10.1007/s12031-018-1155-6
54. Li K, Reichmann H (2016) Role of iron in neurodegenerative diseases. J Neural Transm 123:389–399. https://doi.org/10.1007/s00702-016-1508-7
55. Boddaert N et al (2007) Selective iron chelation in Friedreich ataxia: biologic and clinical implications. Blood 110:401–408. https://doi.org/10.1182/blood-2006-12-065433
56. Koeppen AH, Morral JA, McComb RD, Feustel PJ (2011) The neuropathology of late-onset Friedreich's ataxia. Cerebellum 10:96–103. https://doi.org/10.1007/s12311-010-0235-0
57. Koeppen AH et al (2007) The dentate nucleus in Friedreich's ataxia: the role of iron-responsive proteins. Acta Neuropathol 114:163–173. https://doi.org/10.1007/s00401-007-0220-y
58. Rotig A et al (1997) Aconitase and mitochondrial iron-sulphur protein deficiency in Friedreich ataxia. Nat Genet 17:215–217. https://doi.org/10.1038/ng1097-215
59. Codazzi F et al (2016) Friedreich ataxia-induced pluripotent stem cell-derived neurons show a cellular phenotype that is corrected by a benzamide HDAC inhibitor. Hum Mol Genet 25:4847–4855. https://doi.org/10.1093/hmg/ddw308
60. Auchere F, Santos R, Planamente S, Lesuisse E, Camadro JM (2008) Glutathione-dependent redox status of frataxin-deficient cells in a yeast model of Friedreich's ataxia. Hum Mol Genet 17:2790–2802. https://doi.org/10.1093/hmg/ddn178
61. Wijemanne S, Ondo W (2017) Restless Legs Syndrome: clinical features, diagnosis and a practical approach to management. Pract Neurol 17:444–452. https://doi.org/10.1136/practneurol-2017-001762
62. Earley CJ, Barker PB, Horska A, Allen RP (2006) MRI-determined regional brain iron concentrations in early- and late-onset restless legs syndrome. Sleep Med 7:458–461. https://doi.org/10.1016/j.sleep.2005.11.009
63. Rizzo G et al (2013) Low brain iron content in idiopathic restless legs syndrome patients detected by phase imaging. Mov Disord: Off J Mov Disord Soc 28:1886–1890. https://doi.org/10.1002/mds.25576
64. Haba-Rubio J et al (2005) Restless legs syndrome and low brain iron levels in patients with haemochromatosis. J Neurol Neurosurg Psychiatry 76:1009–1010. https://doi.org/10.1136/jnnp.2003.030536
65. Earley CJ et al (2000) Abnormalities in CSF concentrations of ferritin and transferrin in restless legs syndrome. Neurology 54:1698–1700
66. Ge Y et al (2007) Quantitative assessment of iron accumulation in the deep gray matter of multiple sclerosis by magnetic field correlation imaging. AJNR Am J Neuroradiol 28:1639–1644. https://doi.org/10.3174/ajnr.A0646
67. Haacke EM et al (2009) Characterizing iron deposition in multiple sclerosis lesions using susceptibility weighted imaging. J Magn Reson Imaging: JMRI 29:537–544. https://doi.org/10.1002/jmri.21676
68. Hametner S et al (2013) Iron and neurodegeneration in the multiple sclerosis brain. Ann Neurol 74:848–861. https://doi.org/10.1002/ana.23974
69. Stephenson E, Nathoo N, Mahjoub Y, Dunn JF, Yong VW (2014) Iron in multiple sclerosis: roles in neurodegeneration and repair. Nat Rev Neurol 10:459–468. https://doi.org/10.1038/nrneurol.2014.118
70. Weigel KJ, Lynch SG, LeVine SM (2014) Iron chelation and multiple sclerosis. ASN Neuro 6:e00136. https://doi.org/10.1042/AN20130037
71. Benkler C et al (2010) Recent advances in amyotrophic lateral sclerosis research: perspectives for personalized clinical application. EPMA J 1:343–361. https://doi.org/10.1007/s13167-010-0026-1

72. Kasarskis EJ, Tandon L, Lovell MA, Ehmann WD (1995) Aluminum, calcium, and iron in the spinal cord of patients with sporadic amyotrophic lateral sclerosis using laser microprobe mass spectroscopy: a preliminary study. J Neurol Sci 130:203–208
73. Ignjatovic A et al (2012) Inappropriately chelated iron in the cerebrospinal fluid of amyotrophic lateral sclerosis patients. Amyotroph Lateral Scler 13:357–362. https://doi.org/10.3109/17482968.2012.665929
74. Golko-Perez S, Amit T, Bar-Am O, Youdim MB, Weinreb O (2017) A novel iron chelator-radical scavenger ameliorates motor dysfunction and improves life span and mitochondrial biogenesis in SOD1(G93A) ALS Mice. Neurotox Res 31:230–244. https://doi.org/10.1007/s12640-016-9677-6
75. Li X, Jankovic J, Le W (2011) Iron chelation and neuroprotection in neurodegenerative diseases. J Neural Transm (Vienna) 118:473–477. https://doi.org/10.1007/s00702-010-0518-0
76. Siciliano G et al (2007) Antioxidant capacity and protein oxidation in cerebrospinal fluid of amyotrophic lateral sclerosis. J Neurol 254:575–580. https://doi.org/10.1007/s00415-006-0301-1
77. Goodall EF, Haque MS, Morrison KE (2008) Increased serum ferritin levels in amyotrophic lateral sclerosis (ALS) patients. J Neurol 255:1652–1656. https://doi.org/10.1007/s00415-008-0945-0
78. Mizuno Y et al (2006) Transferrin localizes in Bunina bodies in amyotrophic lateral sclerosis. Acta Neuropathol 112:597–603. https://doi.org/10.1007/s00401-006-0122-4
79. Wang XS et al (2004) Increased incidence of the Hfe mutation in amyotrophic lateral sclerosis and related cellular consequences. J Neurol Sci 227:27–33. https://doi.org/10.1016/j.jns.2004.08.003
80. Perls T (2004) Dementia-free centenarians. Exp Gerontol 39:1587–1593. https://doi.org/10.1016/j.exger.2004.08.015
81. Bush AI (2008) Drug development based on the metals hypothesis of Alzheimer's disease. J Alzheimers Dis 15:223–240
82. Ayton S, Lei P, Bush AI (2015) Biometals and their therapeutic implications in Alzheimer's disease. Neurotherapeutics 12:109–120. https://doi.org/10.1007/s13311-014-0312-z
83. Crapper McLachlan DR et al (1991) Intramuscular desferrioxamine in patients with Alzheimer's disease. Lancet 337:1304–1308
84. Fine JM et al (2012) Intranasal deferoxamine improves performance in radial arm water maze, stabilizes HIF-1alpha, and phosphorylates GSK3beta in P301L tau transgenic mice. Exp Brain Res 219:381–390. https://doi.org/10.1007/s00221-012-3101-0
85. Guo C et al (2013) Intranasal deferoxamine reverses iron-induced memory deficits and inhibits amyloidogenic APP processing in a transgenic mouse model of Alzheimer's disease. Neurobiol Aging 34:562–575. https://doi.org/10.1016/j.neurobiolaging.2012.05.009
86. Guo C et al (2015) Intranasal deferoxamine attenuates synapse loss via up-regulating the P38/HIF-1alpha pathway on the brain of APP/PS1 transgenic mice. Front Aging Neurosci 7:104. https://doi.org/10.3389/fnagi.2015.00104
87. Di Vaira M et al (2004) Clioquinol, a drug for Alzheimer's disease specifically interfering with brain metal metabolism: structural characterization of its zinc(II) and copper(II) complexes. Inorg Chem 43:3795–3797. https://doi.org/10.1021/ic0494051
88. LeVine H 3rd, Ding Q, Walker JA, Voss RS, Augelli-Szafran CE (2009) Clioquinol and other hydroxyquinoline derivatives inhibit Abeta(1-42) oligomer assembly. Neurosci Lett 465:99–103. https://doi.org/10.1016/j.neulet.2009.08.002
89. Mancino AM, Hindo SS, Kochi A, Lim MH (2009) Effects of clioquinol on metal-triggered amyloid-beta aggregation revisited. Inorg Chem 48:9596–9598. https://doi.org/10.1021/ic9014256
90. Stoppelkamp S et al (2011) In vitro modelling of Alzheimer's disease: degeneration and cell death induced by viral delivery of amyloid and tau. Exp Neurol 229:226–237. https://doi.org/10.1016/j.expneurol.2011.01.018

91. Cherny RA et al (2001) Treatment with a copper-zinc chelator markedly and rapidly inhibits beta-amyloid accumulation in Alzheimer's disease transgenic mice. Neuron 30:665–676
92. Grossi C et al (2009) Clioquinol decreases amyloid-beta burden and reduces working memory impairment in a transgenic mouse model of Alzheimer's disease. J Alzheimers Dis 17:423–440. https://doi.org/10.3233/JAD-2009-1063
93. Ritchie CW et al (2003) Metal-protein attenuation with iodochlorhydroxyquin (clioquinol) targeting Abeta amyloid deposition and toxicity in Alzheimer disease: a pilot phase 2 clinical trial. Archives of neurology 60:1685–1691. https://doi.org/10.1001/archneur.60.12.1685/12/1685
94. Adlard PA et al (2008) Rapid restoration of cognition in Alzheimer's transgenic mice with 8-hydroxy quinoline analogs is associated with decreased interstitial Abeta. Neuron 59:43–55. https://doi.org/10.1016/j.neuron.2008.06.018
95. Adlard PA et al (2011) Metal ionophore treatment restores dendritic spine density and synaptic protein levels in a mouse model of Alzheimer's disease. PLoS ONE 6:e17669. https://doi.org/10.1371/journal.pone.0017669
96. Crouch PJ et al (2011) The Alzheimer's therapeutic PBT2 promotes amyloid-beta degradation and GSK3 phosphorylation via a metal chaperone activity. J Neurochem 119:220–230. https://doi.org/10.1111/j.1471-4159.2011.07402.x
97. Lannfelt L et al (2008) Safety, efficacy, and biomarker findings of PBT2 in targeting Abeta as a modifying therapy for Alzheimer's disease: a phase IIa, double-blind, randomised, placebo-controlled trial. Lancet Neurol 7:779–786. https://doi.org/10.1016/S1474-4422(08)70167-4
98. Faux NG et al (2010) PBT2 rapidly improves cognition in Alzheimer's Disease: additional phase II analyses. J Alzheimers Dis 20:509–516. https://doi.org/10.3233/JAD-2010-1390
99. Kwiatkowski A et al (2012) Long-term improvement under deferiprone in a case of neurodegeneration with brain iron accumulation. Parkinsonism Relat Disord 18:110–112. https://doi.org/10.1016/j.parkreldis.2011.06.024
100. Grolez G et al (2015) Ceruloplasmin activity and iron chelation treatment of patients with Parkinson's disease. BMC Neurol 15:74. https://doi.org/10.1186/s12883-015-0331-3
101. Devos D et al (2014) Targeting chelatable iron as a therapeutic modality in Parkinson's disease. Antioxid Redox Signal 21:195–210. https://doi.org/10.1089/ars.2013.5593
102. Lei P et al (2015) Clioquinol rescues Parkinsonism and dementia phenotypes of the tau knockout mouse. Neurobiol Dis 81:168
103. Ayton S et al (2013) Ceruloplasmin dysfunction and therapeutic potential for Parkinson disease. Ann Neurol 73:554–559. https://doi.org/10.1002/ana.23817
104. Bar-Am O et al (2015) Neuroprotective and neurorestorative activities of a novel iron chelator-brain selective monoamine oxidase-A/monoamine oxidase-B inhibitor in animal models of Parkinson's disease and aging. Neurobiol Aging 36:1529–1542. https://doi.org/10.1016/j.neurobiolaging.2014.10.026
105. Gal S, Zheng H, Fridkin M, Youdim MB (2005) Novel multifunctional neuroprotective iron chelator-monoamine oxidase inhibitor drugs for neurodegenerative diseases. In vivo selective brain monoamine oxidase inhibition and prevention of MPTP-induced striatal dopamine depletion. J Neurochem 95:79–88. https://doi.org/10.1111/j.1471-4159.2005.03341.x
106. Zheng H et al (2005) Novel multifunctional neuroprotective iron chelator-monoamine oxidase inhibitor drugs for neurodegenerative diseases: in vitro studies on antioxidant activity, prevention of lipid peroxide formation and monoamine oxidase inhibition. J Neurochem 95:68–78. https://doi.org/10.1111/j.1471-4159.2005.03340.x
107. Shachar DB, Kahana N, Kampel V, Warshawsky A, Youdim MB (2004) Neuroprotection by a novel brain permeable iron chelator, VK-28, against 6-hydroxydopamine lession in rats. Neuropharmacology 46:254–263. https://doi.org/10.1016/S002839080300354X [pii]
108. Kaur D et al (2003) Genetic or pharmacological iron chelation prevents MPTP-induced neurotoxicity in vivo: a novel therapy for Parkinson's disease. Neuron 37:899–909. https://doi.org/10.1016/S0896627303001260 [pii]
109. Hasan TF et al (2018) Diagnosis and management of acute ischemic stroke. Mayo Clin Proc 93:523–538. https://doi.org/10.1016/j.mayocp.2018.02.013

110. Iadecola C, Anrather J (2011) Stroke research at a crossroad: asking the brain for directions. Nat Neurosci 14:1363–1368. https://doi.org/10.1038/nn.2953
111. Selim M (2010) Treatment with the iron chelator, deferoxamine mesylate, alters serum markers of oxidative stress in stroke patients. Transl Stroke Res 1:35–39. https://doi.org/10.1007/s12975-009-0001-0
112. Investigators TR (1996) A randomized trial of tirilazad mesylate in patients with acute stroke (RANTTAS). Stroke: J Cereb Circ 27:1453–1458
113. Haley EC Jr (1998) High-dose tirilazad for acute stroke (RANTTAS II). Stroke: J Cereb Circ 29:1256–1257
114. Hanson LR et al (2009) Intranasal deferoxamine provides increased brain exposure and significant protection in rat ischemic stroke. J Pharmacol Exp Ther 330:679–686. https://doi.org/10.1124/jpet.108.149807
115. Tuo QZ et al (2017) Tau-mediated iron export prevents ferroptotic damage after ischemic stroke. Mol Psychiatry 22:1520–1530. https://doi.org/10.1038/mp.2017.171
116. Jang YG, Ilodigwe D, Macdonald RL (2009) Metaanalysis of tirilazad mesylate in patients with aneurysmal subarachnoid hemorrhage. Neurocrit Care 10:141–147. https://doi.org/10.1007/s12028-008-9147-y
117. Burk K (2017) Friedreich Ataxia: current status and future prospects. Cerebellum & Ataxias 4. https://doi.org/10.1186/s40673-017-0062-x
118. Voncken M, Ioannou P, Delatycki MB (2004) Friedreich ataxia-update on pathogenesis and possible therapies. Neurogenetics 5:1–8. https://doi.org/10.1007/s10048-003-0170-z
119. Pandolfo M et al (2014) Deferiprone in Friedreich ataxia: a 6-month randomized controlled trial. Ann Neurol 76:509–521. https://doi.org/10.1002/ana.24248
120. Velasco-Sanchez D et al (2011) Combined therapy with idebenone and deferiprone in patients with Friedreich's ataxia. Cerebellum 10:1–8. https://doi.org/10.1007/s12311-010-0212-7
121. Earley CJ, Heckler D, Allen RP (2004) The treatment of restless legs syndrome with intravenous iron dextran. Sleep Med 5:231–235. https://doi.org/10.1016/j.sleep.2004.03.002
122. Earley CJ et al (2009) A randomized, double-blind, placebo-controlled trial of intravenous iron sucrose in restless legs syndrome. Sleep Med 10:206–211. https://doi.org/10.1016/j.sleep.2007.12.006
123. Pedchenko TV, LeVine SM (1998) Desferrioxamine suppresses experimental allergic encephalomyelitis induced by MBP in SJL mice. J Neuroimmunol 84:188–197
124. Lynch SG, Peters K, LeVine SM (1996) Desferrioxamine in chronic progressive multiple sclerosis: a pilot study. Mult Scler 2:157–160. https://doi.org/10.1177/135245859600200306
125. Lynch SG, Fonseca T, LeVine SM (2000) A multiple course trial of desferrioxamine in chronic progressive multiple sclerosis. Cell Mol Biol (Noisy-le-grand) 46:865–869
126. Creange A, Lefaucheur JP, Balleyguier MO, Galacteros F (2013) Iron depletion induced by bloodletting and followed by rhEPO administration as a therapeutic strategy in progressive multiple sclerosis: a pilot, open-label study with neurophysiological measurements. Neurophysiol Clin = Clin Neurophysiol 43:303–312. https://doi.org/10.1016/j.neucli.2013.09.004

The manufacturer's authorised representative in the EU is Springer Nature Customer Service Centre GmbH, Europaplatz 3, 69115 Heidelberg, Germany. If you have any concerns regarding our products, please contact ProductSafety@springernature.com

Printed and bound by CPI Group (UK) Ltd, Croydon, CR0 4YY
01/08/2026
02183522-0002